ADVANCES IN COGNITIVE RADIO SYSTEMS

Edited by **Cheng-Xiang Wang** and
Joseph Mitola III

Advances in Cognitive Radio Systems
Edited by Cheng-Xiang Wang and Joseph Mitola III

Published by InTech
Janeza Trdine 9, 51000 Rijeka, Croatia

Notice

Statements and opinions expressed in the chapters are these of the individual contributors and not necessarily those of the editors or publisher. No responsibility is accepted for the accuracy of information contained in the published chapters. The publisher assumes no responsibility for any damage or injury to persons or property arising out of the use of any materials, instructions, methods or ideas contained in the book.

Publishing Process Manager Bojan Rafaj

Technical Editor Teodora Smiljanic

Cover Designer InTech Design Team

Additional hard copies can be obtained from orders@intechopen.com

Advances in Cognitive Radio Systems, Edited by Cheng-Xiang Wang and Joseph Mitola III
p. cm.
ISBN 978-953-51-0666-1

For EU product safety concerns:
IN TECH d.o.o., Prolaz Marije Krucifikse Kozulić 3, 51000 Rijeka, Croatia,
info@intechopen.com or visit our website at intechopen.com.

Contents

1

Wideband Voltage Controlled Oscillators for Cognitive Radio Systems

Alessandro Acampora and Apostolos Georgiadis
Centre Tecnològic de Telecomunicacions de Catalunya (CTTC)
Spain

1. Introduction

In the latest years much research effort was devoted to envision a new paradigm for wireless transmission. Results from recent works (Wireless Word Research Forum, 2005) indicate that a possible solution would lie in utilizing in a more efficient manner the diverse Radio Access Technologies[1] (RATs) that are available nowadays, with the purpose of enabling interoperability among them and convergence into one global telecom infrastructure (beyond 3G).

Turning such a representation into reality requires endowing both the network and the user terminal with advanced management functionalities to ensure an effective utilization of radio resources. From the network providers' side, this translates in devising support for heterogeneous RATs, to map or reallocate traffic stream according to QoS requirements[2], while from the users terminals' side a major step towards a smarter utilization of radio resources consists in enabling reconfigurability, so to adapt dynamically the transmission to the spectrum environment in such a way that is no longer required to have fixed frequency bands mapped uniquely to specific RAT. Through a smarter selection of unused frequency bands spanning various access technologies, is possible to achieve the maximization of each RAT capacity both in time and space (within a geographical area) while at the same time minimizing the mutual interference. The support for heterogeneous access technologies on the network side and reconfigurable devices on the terminal side constitutes the essence of the Cognitive Radio paradigm (Akyildiz et al., 2006).

Several spectrum management protocols have been proposed from different research bodies/agencies worldwide, e.g. DARPA XG OSA "Open Spectrum Access" in (Akyildiz et al., 2006). However, all of them pose relevant challenges from the hardware implementation point of view to achieve adaptive utilization of radio resources. In fact, in order to identify unused portion of the spectrum at a specific time in a certain geographical area is necessary

---

[1]Consider for example GSM/GPRS for 2G cellular network, UMTS (HSPA) for 3G (3.5G) cellular network delivering high speed data transmission and nomadic internet access, WLAN for wireless local area networks, WIMAX for providing wireless metropolitan internet access.
[2]Examples of protocols offering support for managing heterogeneous networks are GAN "Generic Access Network" and ANDSF "Access Network Discovery and Selection Function", details can be found in (Ferrus et al., 2010; Frei et al., 2011)

to execute a real-time, wide-band sensing, capable of spanning across the frequency bands of the various RATs. To that aim the frequency of the local oscillator in the transceiver module of a user terminal should be continuously swept across a wide frequency range, thus motivating the need for wideband tunable oscillators as an enabling technology for successful deployment of Cognitive Radio capabilities. There are many possibilities to implement an oscillator with a variable frequency, the most common of which is referred to Voltage Controlled Oscillator (VCO) in which generally altering a DC voltage at a convenient node in the circuit produces a frequency shifting in the sinusoidal output waveform. In the case of VCOs derived by harmonic oscillators[3] this could be due the variation in the parameters of the nonlinear device model (Sun, 1972), or simply the effect of an added varactor in the embedded fixed frequency oscillator network (Cohen, 1979; Peterson, 1980) so that the phase of the signal across the feedback path could be varied, and the its frequency adjusted as a result of a variable capacitive loading.

A suitable VCO for Cognitive Radio applications should provide large tuning bandwidths in order to cover the spectrum of the diverse RATs, and has to cope with additional limitations due to space occupancy of the circuit (the possibility of having an integrated chip), its spectral purity (expressed in terms of low phase noise), the linearity of the tuning function, its harmonic rejection (related to the content of higher order harmonics with respect to the fundamental) its output power (which is assumed to be as high as possible) its efficiency (the amount of RF energy output produced relative to the DC power supply) and DC current consumption (which ideally should be kept low). Meeting all these requirements might be made easier if instead of using conventional circuit techniques, one considers microwave distributed voltage controlled oscillators (DVCO) (Divina & Škvor 1998; Wu & Hajimiri, 2000; Yuen & Tsang, 2004).

Essentially a distributed oscillator consists of a distributed amplifier (Škvor et al., 1992; Wong, 1993) in which a feedback path is created in order to build up and sustain oscillations. In order to vary the oscillation frequency in a prescribed range is possible to introduce a varactor in the feedback loop (Yuen & Tsang, 2004) or use some advanced techniques like the "current steering" in (Wu & Hajimiri, 2000). However, these solutions do not provide a real wide-band operation since relative tuning ranges of nearly 12% are attained both in (Yuen & Tsang, 2004) and in (Wu & Hajimiri, 2000). Instead the reverse mode DVCO working principle (Divina & Škvor, 1998; Škvor et al., 1992) based upon a feedback path for backward scattered waves in the drain line (hence the name) and the concurrent variation the active devices' gate voltages as a mean for adjusting the oscillation frequency, presents a wide tuning range, up to a frequency decade (Škvor et al., 1992) a good output power, on the order of +10 dBm, adequate suppression of higher harmonics with typical values for second and third order harmonic rejection of -20 dBc, -30dBc respectively, and a satisfactory spectral purity, with an average phase noise on the order of -100 dBc/Hz at 1 MHz offset from the carrier across the 1 GHz tuning bandwidth in (Acampora et al., 2010), allowing for fine spectral resolution. Yet, it suffers from a major drawback, which resides in its tuning function, i.e. the variation of oscillation frequency

[3]This is not the only option. In the case of digital IC for example, the VCOs are based on *relaxation oscillators* (ring oscillators, delay line oscillators, rotary travelling wave oscillators) which using logical gates synthesize square, triangular, sawtooth waveforms as for example in (Zhou et al., 2011).

with respect to the control voltages, which in the case of large signal operation sensibly deviates from linear analysis prediction. In (Divina & Ŝkvor 1998), small signal analysis techniques were used to model the DVCO behaviour, explaining the basic mechanism for which tuning is made possible, which consists in opportunely altering the phase characteristic of the DVCO by changing the transconductance of the active devices through their gate bias voltages. This approach, although analytical, it is limited in that it doesn't allow one to identify important oscillator figures of merit (e.g. oscillation power level, higher order harmonics content, and oscillation's stability) since it only detects the frequency at which oscillations build-up. In order to cope with these issues, nonlinear simulation techniques must be employed in the Time Domain (TD) (Silverberg. & Wing, 1968; Sobhy & Jastrzebski 1985) in the Frequency Domain (FD) (Rizzoli et al., 1992) or in a "mixed" Time-Frequency domain (Ngoya & Larcheveque, 1996). In TD simulations, the differential system of equation is numerically integrated with respect to the time variable, delivering the most accurate representation of the solution waveform, which enables the transient[4] and the steady state analysis as well. In FD simulations the circuit variables are conveniently expressed in terms of generalized Fourier series[5], which permits to quickly have information about the steady state, skipping the transient evaluation (Kundert et al., 1990). Application of this principle in microwave circuit analysis gave rise to the Harmonic Balance (HB) method (Rizzoli & Neri, 1988; Rizzoli et al., 1992) and its extension to modulated signals, the Envelope Transient simulation (Brachtendorf et al., 1998; Ngoya et al., 1995, Ngoya & Larcheveque, 1996) which is a mixed TD/FD method.

The authors in (Divina & Škvor, 1998) make use of TD simulations to assess the nonlinear oscillator performance. However, oscillator transient simulations are very time-consuming since many cycles of an high frequency carrier have to be waited out, until the transient is extinguished and the steady state is settled down (Giannini & Leuzzi, 2004). Furthermore, in the case of the DVCO, transient simulations are often prone to numerical instabilities and convergence failure due to the time domain evaluation of distributed elements which are frequency dispersive (Suarez & Quèrè, 2003). When analyzing a multi-resonant distributed microwave circuit, with multiple oscillation modes like the DVCO (Acampora et al., 2010; Collado et al., 2010) this issue turns out to be particularly undesireable.

For the aforementioned reasons, in this work the reverse mode DVCO tuning function is calculated by employing HB simulation techniques, opportunely modified to take into account the autonomous nature of the circuit being studied. In fact, an HB simulation of an oscillator circuit is prone to errors like convergence failure or convergence to DC equilibrium point ("zero frequency solution") since it is not externally driven by time-varying RF generators (Chang et al., 1991). Probe methods aim at eliminating the ambiguity by having a fictitious voltage sine-wave RF generator with unknown amplitude and

[4] This is the reason why often the terms "Time Domain simulation" and "Transient simulation" are often used interchangeably.

[5] Assuming a periodic or quasi-periodic solution exists, it will retain all the features of the RF generators acting as sources. In particular, if ($\omega_1, \omega_2, \ldots, \omega_k, \ldots \omega_n$) are the n input incommensurable frequencies of the sources, a general circuit variable will contain intermodulation products ($p_1\omega_1+p_2\omega_2+\ldots+ p_k\omega_k +\ldots+ p_n\omega_n$) where p_i are integer coefficients. See (Kundert, 1997, 1999) for more details.

frequency (its phase is conveniently set to zero) inserted at an appropriate node in the circuit in order to force the HB simulator to converge to the oscillating solution. Both amplitude and probe's frequency represent two extra variables which are found by imposing a non-perturbation condition at the node in the circuit to which the probe is connected.

Using these techniques, a reverse mode DVCO has been successfully designed and implemented using standard prototyping techniques and off-the-shelf inexpensive components. The topology of the DVCO resembled a feedack distributed amplifier having four sections and employing a NE 3509M04 HJ-FETs as active elements providing the necessary gain for triggering oscillations. The inter-sections coupling network consited in π-type m-derived sections which comprised lumped inductors and capacitor, the input/output parasitic capacitance of each FET and microstrip line sections providing interconnections and access to each device from the drain line/gate line, and behaved as a low pass structure (Wong, 1993), with a nominal impedance of 50 Ω and a cutoff frequency of 3 GHz. Experimental plots revealed a reduction in the frequency tuning range (0.75−1.85 GHz) with respect to the simulated one (1−2.4 GHz), but still assuring a wideband operation (delivering an 85% relative tuning range). Phase noise measurements were performed to validate the effectiveness of the proposed DVCO for practical purposes, obtaining a mean value of -111.2 dBc/Hz at 1 MHz offset from the carrier, across the overall tuning range. Measured Output Power level was comprised between +5 dBm and +7.5 dBm.

The chapter is organized as follows. In section 2 the distributed amplifier/oscillator/ VCO working principle is introduced and some examples of its implementation will be given. In section 3 the necessary background in TD/FD simulation techniques is provided with particular emphasis to HB balance/ probe methods for oscillator analysis. Section 4 deals with the analysis and design of a four-section distributed voltage controlled oscillator. Section 5 is devoted to the implementation details and measurements. Last section concludes the work, and paves the way for future research.

2. Distributed voltage controlled oscillator linear analysis

This section is aimed at understanding the working principle of Distributed Microwave Amplifiers and Oscillators/VCO.

2.1 Introduction – Distributed amplifier and oscillator

In recent years, renewed interest towards distributed microwave circuits has been shown. New architectures for mixers (Safarian et al., 2005), Low Noise Amplifiers (LNA) (Heydary, 2005) oscillators and VCO (Divina & Škvor, 1998; Wu & Hajimiri, 2001), have been proposed, and all of them are susceptible to be implemented in integrated form.

Although highly appreciated today, all these circuits share an old discovery patented by Percival in 1937 (Percival, 1937) and later on published by Gintzon (Gintzon et al., 1948) called "distributed amplification". In his work was explained for the first time how to design a very wideband amplifier provided that several active devices should be used. It turned out that the utilization of a pair artificial k-constant transmission line periodically coupled by the active devices' transconductance (Wong, 1993; Pozar, 2004) provided to the overall structure a linear increase in gain and a very wideband operation. The rationale behind this improvement lied in the fact that artificial transmission line (ATL) sections, made out of lumped inductances and capacitances, were valuable in diminishing the values of parasitic capacitance seen at the output of a single stage, improving noticeably the bandwidth performance.

In Fig. 1 is depicted a three section distributed amplifier. The signal is injected through the gate line (input ATL) and as the travelling waves pass through each section, it gets amplified at the drain line (output ATL) in a concurrent way. At the end of the gate and at the beginning of the drain line a matched termination section (indicated as a resistor), having the same impedance of the ATL is introduced with the purpose of absorbing the forward propagating waves in the gate line and the backward propagating waves in the drain line (Pozar, 2004). A broadband impedance matching network is placed halfway among the last sections to adjust the impedance levels in order to avoid signal reflections. As the operating frequency increases, the lumped elements should be substituted by properly designed transmission line sections. From this arrangement, a distributed oscillator can be obtained introducing a feedback loop between the output line and the input line of the distributed amplifier (Fig. 2). This topology is known as forward gain distributed oscillator, since it involves forward propagating waves that, circulating in the feedback loop, are re-inserted in the input line through the output node. The feedback path length determines the operating frequency; as the path length gets smaller, the maximum attainable frequency increases. Restricted tuning capabilities can be incorporated in this circuit introducing a varactor diode in the feedback line in order to control its electrical length changing the capacitive loading (Yuen & Tsang, 2004), or by adequately modifying the bias currents of the active devices to provide "current-steering delay balanced" tuning in (Wu & Hajimiri, 2001).

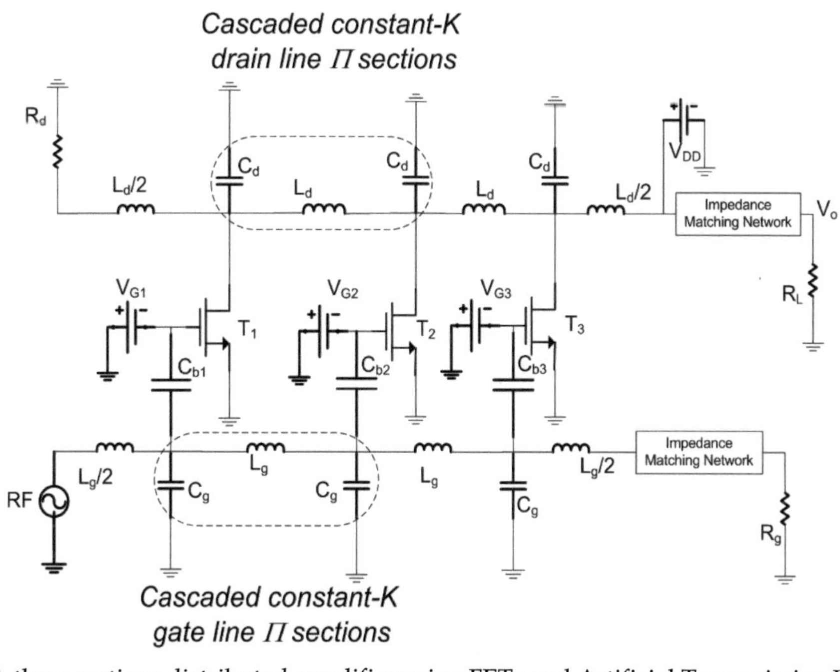

Fig. 1. A three sections distributed amplifier using FETs and Artificial Transmission Lines.

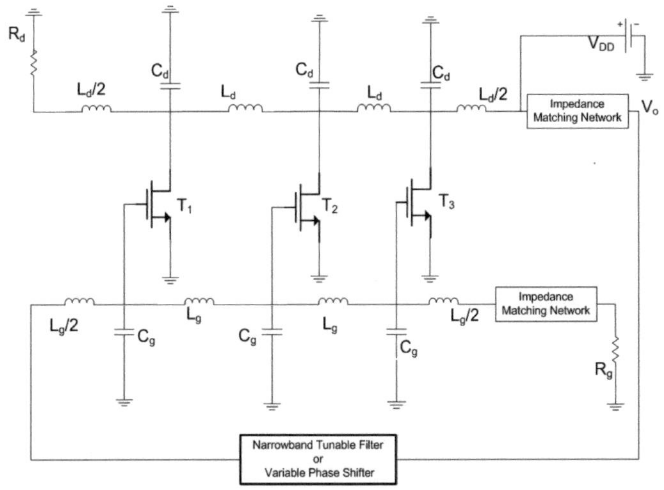

Fig. 2. An ideal DVCO, with a tuning element in the feedback loop.

2.2 Reverse gain mode distributed voltage controlled oscillator

An alternative topology for the distributed oscillator was proposed by Škvor (Škvor et al., 1992). The possibility of removing the dummy drain resistor and connecting together the drain and the gate lines in a "reverse manner" was contemplated (Fig. 3) in order to exploit the "backward" scattered waves in the drain line, making them available once more through a feedback loop to the gate line. Compared to the high frequency DVCO proposed in (Wu & Hajimiri, 2001) it offers several advantages, mainly in terms of greater output power and wider tuning bandwidth (Divina & Škvor, 1998), at the expense of added complexity residing in the tuning algorithm, which should be fully analyzed employing nonlinear numerical techniques.

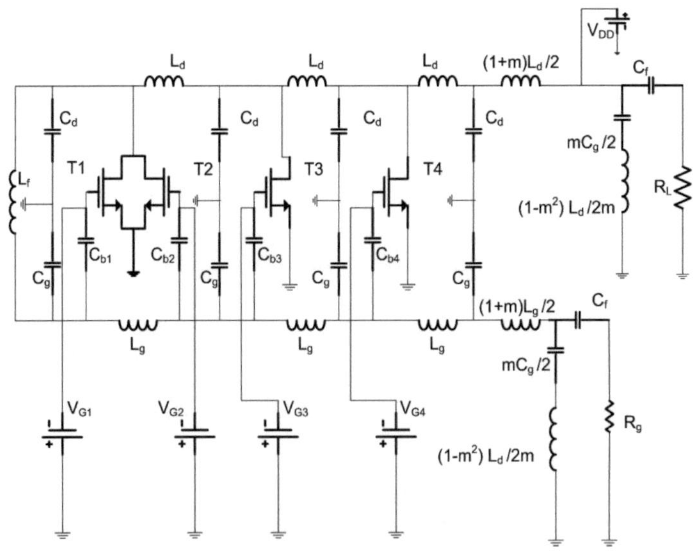

Fig. 3. Idealized Schematic of the Reverse Gain DVCO.

In this circuit, the effective feedback path length seen by the microwave signals can be modified by activating only one active device at a time, leaving the others switched off. As a consequence, considering a DVCO with N sections, a set of N discrete oscillations will be produced, whose frequencies are distributed within the pass-band of the ATLs (constant k-filter sections (Divina & Škvor, 1998)). These frequencies have a precise relationship with the cut-off frequency of the LC cells in such a way that the highest frequency component will correspond to the activation of the first transistor, and the lowest frequency will be obtained with the activation of the last active device (Fig.3). Oscillation Frequencies are seen in decreasing order as we subsequently activate each stage, from the first to the last. To estimate them analytically, Barckhausen-Nyquist criteria can be applied in the first place so to find the frequency at which the closed loop gain transfer function equals one, permitting signal regeneration[6]. When the p-th stage is activated, oscillation start-up depends on the ratio between the input and the output voltage wave at the p-th stage[7] (called *reverse gain)* which in turn is influenced by the artificial transmission line impedance $Z^{\pi}(\omega)$, the transconductance of the device itself $g_{m(p)}$, and the phase of the signal across the path from the drain line back to the gate line $\Phi_{rev}^{(p)}(\omega)$ (Divina & Škvor, 1995):

$$G_{rev}^{(p)}(\omega)=\frac{g_{m(p)}Z_F(\omega)}{2}e^{-j[\pi+(2p-1)\psi_F(\omega)]}=\left|G_{rev}^{(p)}\right|(\omega)e^{j\Phi_{rev}^{(p)}(\omega)} \tag{1}$$

being:

$$Z^{\pi}(\omega)=Z_F(\omega)e^{j\psi_F(\omega)} \tag{2}$$

the impedance of the π-type LC sections. Applying Nyquist Criterion for the onset of oscillations to $G_{rev}^{(p)}(\omega)$, a relation that express the possible self resonant frequencies as a function of the active device position p is obtained (Škvor et al., 1992):

$$\frac{f^{(p)}}{f_c}=\sin\left(\frac{\pi}{4p-2}\right) \tag{3}$$

where f_c is the cut-off frequency of the low pass structures.

Device Switched ON (p)	Oscillation frequencies *(Analytical linear model)*
T_1	4 GHz
T_2	2 GHz
T_3	1.24 GHz
T_4	880 MHz

Table 1. DVCO frequencies for an embedded ATL cut-off frequency of 4 GHz.

[6] Barckhausen-Nyquist criteria states that for a feedback amplifier with a gain $A(\omega)$ and a feedback transfer function $\beta(j\omega)$, an oscillation occurs at the frequency such that the closed $|A(\omega_0)\beta(j\omega_0)|=1$, $\measuredangle(A(\omega_0)\beta(j\omega_0))=0$.

[7] In the case of simplified linear (small signal) analysis, in which are neglected all the parasitics of the transistors that are simply modeled as controlled current sources, and it is assumed to use the same values for the lumped elements both in the drain an in the gate line. Frequency dependent part only accounts for the impedance of the constant k- sections.

As an example, in a four sections DVCO with a specified cut-off frequency of 4 GHz, the (3) provides the frequencies given in Tab. 1.

Apart from generating discrete signals, this circuit possess appealing tuning capabilities. In fact, if two stages are activated at the same time, by proper regulation of the biasing voltages[8] it is possible to get a whole range of frequencies which fall in between the two discrete frequencies related to the activation of each single active device. Its operation principle could be explored analytically, by considering the reverse gain of p-th and of q-th ($q > p$) stages (Divina & Škvor, 1995) when both are simultaneously active:

$$G_{rev}^{(p,q)}(\omega)=\sqrt{\left|G_{rev}^{(p)}(\omega)\right|^2+2\left|G_{rev}^{(p)}(\omega)\right|\left|G_{rev}^{(q)}(\omega)\right|\cos\left(\phi_p-\phi_q\right)+\left|G_{rev}^{(q)}(\omega)\right|^2}$$
$$\tan\left(\Delta\Phi_{rev}^{(p,q)}(\omega)\right)=\frac{\left|G_{rev}^{(p)}(\omega)\right|\sin\left(\phi_p\right)+\left|G_{rev}^{(q)}(\omega)\right|\sin\left(\phi_q\right)}{\left|G_{rev}^{(p)}(\omega)\right|\cos\left(\phi_p\right)+\left|G_{rev}^{(q)}(\omega)\right|\cos\left(\phi_q\right)} \tag{4}$$

Introducing the expression for the reverse gains (5) and simplifying the expression for the phase in (4) one gets:

$$G_{rev}^{(p)}(\omega)=\frac{g_{m(p)}Z^{\pi}(\omega)}{2}e^{-j[\pi+(2p-1)\psi_F(\omega)]}$$
$$G_{rev}^{(q)}(\omega)=\frac{g_{m(q)}Z^{\pi}(\omega)}{2}e^{-j[\pi+(2q-1)\psi_F(\omega)]} \tag{5}$$

$$G_{rev}^{(p,q)}(\omega)=\frac{Z^{\pi}(\omega)}{2}\sqrt{g_{m(p)}^2+2g_{m(p)}\cdot g_{m(q)}\cos\left[(q-p)\psi_F(\omega)\right]+g_{m(q)}^2}$$
$$\Delta\Phi_{rev}^{(p,q)}(\omega)=-\frac{[\pi+(2p-1)\psi_F(\omega)]g_{mp}+[\pi+(2q-1)\psi_F(\omega)]g_{mq}}{g_{mp}+g_{mq}} \tag{6}$$

which illustrate that the phase of the reverse gain can be changed by means of the trans-conductance of the active devices p and q. This phase variation is the responsible for the frequency tuning in the range $[f_q, f_p]$. The best frequency tuning strategy turns out to be based on the complementary biasing of two adjacent active devices ($q = p+1$). The gate voltage of the p-th section is decreased while simultaneously the gate voltage of the (p+1)-th section is increased, in order to tune the frequency from f_p down to f_{p+1}. This process is applied between each pair of transistors to get a tuning range from f_N to f_1 for a DVCO with N sections. In (Škvor et al., 1992) it has been shown that it is necessary to insert an additional transistor, placed "crosswise" between the first and the second active stages, in order to provide a supplementary trans-conductance and consequently achieve a continuous phase variation, assuring smooth tuning. Moreover possible spurious oscillations are avoided by matching terminal sections effectively (with a gate/drain line reflection coefficient not greater than -20 dB ideally) to the gate and to the output loads (Divina & Škvor, 1995) which

[8]In a common source amplifier, we could vary the input voltage in the gate or the output voltage in the drain to alter the bias of the active devices. The former way is preferred, so when we mention "tuning voltages" or "tuning controls" we will refer to the variation of the gate voltages of the FET devices.

is accomplished placing an *m*-derived section, just before the drain/gate output terminations. For a value of m=0.6 a broadband matching is assured over nearly the 85% of the band of interest (Wong, 1993).

2.3 Linear analysis of the DVCO and preliminary design

A four section reverse mode DVCO can be designed starting from the linear analysis that has been hitherto shown (Acampora et al., 2010; Collado et al., 2010). A simplified DVCO schematic is set up in a commercial microwave circuits' simulator and contains ideal DC blocks and DC feeding networks and resistors in the drain, which are deemed necessary for proper biasing in the real case aren't optimized. In this idealized situation the values of inductances and capacitances chosen for the artificial transmission line sections, are such that the cut-off frequencies both in the gate and in the drain line were of 3 GHz ($f_c^{(g)}=f_c^{(d)}$) and the characteristic impedance 50 Ω providing: L_g (L_d) =5.3 nH and C_g (C_d)=2.1 pF. Subsequently, are introduced the four HJ-FET nonlinear models for the NE 3509M04. Each section is analysed with the help of S-parameters, in order to have to have minimum phase mismatch between gate/drain line signal propagation (phase balance). To that aim, several design iteration are necessary to optimize the values of inductances and capacitance to use. Finally, the ideal components are substituted by real ones, introducing the layout elements (microstrips lines, cross, bends and tees), the vendor models for the capacitances and inductances, the interconnecting pads and modelling the parasitics due to the insertion of via holes. Taking into account the new layout constraints the values of the lumped components might be re-tuned several times. In (Acampora et al., 2010) the values L_g (L_d)=3 nH and C_g (C_d)=1.2 pF were eventually chosen, providing a cutoff frequency of 5 GHz and a 50 Ω impedance for the basic artificial transmission line sections; the matching *m-derived* sections are then designed accordingly[9].

In order to get an estimate of the potential oscillation frequencies a preliminary linear analysis is performed. To that aim, the small signal admittance $Y_p(f) = G_p(f)+jB_p(f)=$ $\mathrm{Re}(Y_p(f))+j\mathrm{Im}(Y_p(f))$ *at a convenient node P in the circuit* needs to be evaluated and, in order to have a DC unstable solution (Giannini & Leuzzi, 2004) frequency regions for which hold $G_p(f)<0$; $\partial B_p(f)/\partial f > 0$ should be sought for different values of the bias voltages. To that aim a small signal voltage probe[10] is introduced at the node *P* and a frequency swept AC analysis is carried out, measuring the real and imaginary part admittance function $Y_p(f, V_b, \xi_1,\ldots, \xi_k) = I_p/V_p$, where V_b represent the bias voltage and $(\xi_1,\ldots, \xi_k)$ a set of adjustable parameters which can be varied to meet the specifications. The graphs are then displayed and by visual inspection are found the regions in which the admittance real part (conductance) becomes negative presenting "valleys" and its imaginary part (susceptance) presents positive slope. These frequency zones represent unstable DC solution points at

---

[9]The design of the *k- constant LC section* is performed, via the formulas $Z_0=\sqrt{(L/C)}$, $\omega_c=(2/\sqrt{(LC)})$ so once the nominal impedance and the cut-off frequency are chosen, *L* and *C* are uniquely determined. The *m-derived section* inductance and capacitance values, are related to those used in the *k- constant LC section*, having them multiplied by a scale factor. For m=0.6, those values are $C_m=(3/10)C$, $L_m=(3/10)L$, $L_p=(8/15)L$, see (Wong, 1993; Pozar, 2004).

[10]If the probing voltage needs to cause only slight variation around the transistors' quiescient point, a peak value of 10 mV can be considered „small"when the bias voltages are on the order of 1.5–2V.

which the circuit will likely oscillate. When the two conditions stated above hold, a good estimate of the oscillation frequencies is given by the susceptance intercept of the frequency axis, i.e. the points at which $B_p(f)=0$. This linear analysis has shown that potential oscillations could occur along the complete desired frequency band. By turning separately each active device on, a set of four discrete oscillations is obtained, whose estimated frequencies fall within the band [1.2 – 2.6 GHz]. Tuning is achieved continuously since negative conductance zone overlap themselves (Acampora et al., 2010, Collado et al., 2010). However, the linear analysis doesn't provide any information about the steady state oscillation power since it only estimates its frequency, and this approximation could be poor cause of nonlinear operation mode.

3. Nonlinear simulation techniques for microwave oscillators

DVCOs have traditionally been studied and designed using linear design tools (Divina & Škvor, 1998; Wu & Hajimiri, 2001). However, in order to have a realistic picture that could take into account possible instabilities, waveform distortion, optimal power design, phase noise analysis, harmonic content and other relevant performance parameters, one should have recourse to nonlinear simulation techniques (Kundert, 1999). Taking advantage of the latter it is possible to obtain the DVCO tuning curves, which express the values of the gate voltages necessary to synthesize each of the desired frequencies in a prescribed range, the DVCO power level in across the tuning range and the harmonic rejection (Acampora et al., 2010).

The DVCO behaviour could be analyzed numerically integrating the system of differential equations governing the circuit directly in the time domain (Sobhy & Jastrzebski, 1985), having a complete representation of the solution throughout an observation range. Nevertheless, this approach presents severe drawbacks in the case of an oscillator for its long computation times, since many periods of a high frequency carrier need to be evaluated until the steady state is finally reached (Giannini & Leuzzi, 2004). In case of employing microwave distributed elements like in the case of the DVCO, additional processing power is required for representing them in the time domain for they are frequency dispersive; numerical formulation is thus complicated by the introduction of a convolution integral for taking into account this effect. In a DVCO a time domain analysis might be a very frustrating task, considering the possibility of having multi-mode multiple oscillation frequencies which should be manually checked for every particular bias configuration.

A different approach consists in avoid solving the equations in the time domain, but rather use a Fourier series expansion which allow one to transform the original problem into a simpler one, in the frequency domain. This way the entire network is subdivided into two parts; one containing linear microwave devices (both lumped and distributed) which are simply depicted in frequency domain by means of their transfer functions[11], the other containing nonlinearities for which the time domain description is kept, and Discrete Fourier Transform algorithms are used to switch from one domain to the other. A current

[11] Instead of evaluating numerically a complicated convolution integral in time domain, a multiplication of complex quantities is performed, which sensibly relieves the computational load.

balance via the KCL is then established between the two subsets in the form of nonlinear algebraic system of equations involving the unknown coefficients of the Fourier Series Expansion (harmonics), which could be solved by direct or iterative methods (Suarez & Quéré, 2003). This is the essence of the Piecewise Harmonic Balance (HB), which undoubtedly presents the advantage of detecting the periodic steady state solution avoiding the computation of the initial transients.

3.1 Harmonic balance for periodically driven microwave circuits

The starting point for the description of the method (Kundert et al., 1990; Rizzoli et al., 1992; Suarez & Quéré, 2003) is to split the network in the linear part and in the part containing nonlinear devices; the connection among the two parts is provided at q ports. Nonlinear devices are then represented as nonlinear controlled sources, being dependent upon a set of variables called state variables, which are essential in describing the time evolution of the system. No assumption is made upon the nature of the state variables and on the nonlinearities which can be voltages, currents, fluxes or charges. Lastly, the action of independent sources must be taken into account. A set of three vectors is thus considered, being $\mathbf{x}(t)$ the state variable vector, $\mathbf{y}(t)$ the vector containing the nonlinear controlled sources and the generators $\mathbf{g}(t)$:

$$\begin{aligned}\mathbf{x}(t) &= (x_1(t), x_2(t), \cdots, x_n(t))^{\mathrm{T}} \\ \mathbf{y}(t) &= (y_1(t), y_2(t), \cdots, y_q(t))^{\mathrm{T}} \\ \mathbf{g}(t) &= (g_1(t), g_2(t), \cdots, g_s(t))^{\mathrm{T}}\end{aligned} \tag{7}$$

and the most general relationship between $\mathbf{y}(t)$ and $\mathbf{x}(t)$ is of the following form:

$$\mathbf{y}(t) = \mathbf{\Psi}\left(\mathbf{x}(t),\ \mathbf{x}(t-\tau),\ \frac{d\mathbf{x}}{dt},\ \cdots,\ \frac{d^m\mathbf{x}}{dt^m}\right) \tag{8}$$

being $\mathbf{\Psi}(\cdot)$ a *nonlinear function of* $\mathbf{x}(t)$, which generally accounts for a dependence upon shifted values of $\mathbf{x}(t)$, and on its derivatives up to m-th order. In the following, the case for a single generator (*single tone analysis*) is described, so the last vector in (7) reduces to a scalar function $g(t)$=$G_s\sin(\omega t)$; generalizations to multi-tone analysis can be found in (Giannini & Leuzzi, 2004; Suarez & Quéré, 2003). The second step consists in representing all this variables in a generalized Fourier basis of complex exponentials tones. If the generator drives the circuit with angular frequency ω:

$$\mathbf{x}(t) = \sum_{p=-N_m}^{N_m} \mathbf{X}_p e^{jp\omega t},\ \mathbf{y}(t) = \sum_{p=-N_m}^{N_m} \mathbf{Y}_p e^{jp\omega t},\ g(t) = G_s\sin(\omega t) \tag{9}$$

where only a finite number of harmonics N_m has been considered in the expansion. This way the problem is transformed into the frequency domain and the objective becomes the determination of the harmonic components of the two sets of vectors[12] $\mathbf{X}(\omega)$, $\mathbf{Y}(\omega)$ with

[12]Fourier coefficients of a real valued function possess *Hermitian symmetry*, i.e. a series coefficient evaluated at a negative index (-k) is the complex conjugate of the same coefficient computed in its

$\mathbf{X}(\omega)=[\mathbf{X}_{-Nm},\ldots, \mathbf{X}_{h},\ldots, \mathbf{X}_{Nm}]$, $\mathbf{Y}(\omega)=[\mathbf{Y}_{-Nm},\ldots, \mathbf{Y}_{h},\ldots, \mathbf{Y}_{Nm}]$ being each column the *h-harmonic component* ($-N_m \leq h \leq N_m$) of the state variables vector and of the nonlinear device outputs. Since nonlinearities don't admit a frequency representation in terms of transfer functions, a Discrete Fourier Transform (indicated with $\mathcal{F}$) is employed to toggle from time domain samples to the frequency domain:

$$\mathbf{X}(\omega)\xrightarrow{\mathcal{F}^{-1}}\mathbf{x}(t);\ \mathbf{x}(t)\xrightarrow{\Psi(\mathbf{x}(t))}\mathbf{y}(t)\ ;\ \mathbf{y}(t)\xrightarrow{\mathcal{F}}\mathbf{Y}(\omega)=\mathcal{F}\left(\mathbf{y}\left(\mathcal{F}^{-1}(\mathbf{X}(\omega))\right)\right)\Rightarrow \\ \mathbf{Y}(\omega)=\mathbf{Y}(\mathbf{X}(\omega)) \quad (10)$$

Finally, Kirchhoff Current Law equations are written, balancing the harmonic contributions coming from the vectors $\mathbf{X}(\omega)$, $\mathbf{Y}(\omega)$, and from the driving term:

$$\mathbf{H}(\mathbf{X}(\omega))=\left[\mathbf{A}(\omega)\right]\cdot\mathbf{X}(\omega)-\left[\mathbf{B}(\omega)\right]\cdot\mathbf{Y}(\mathbf{X}(\omega))-c\cdot G_s\mathbf{e}_k=\mathbf{0} \quad (11)$$

which takes the form of a linear relationship between these three vectors (Rizzoli et al., 1992; Suarez & Quéré, 2003; Giannini & Leuzzi , 2004), relating node voltages to branch currents and in which $[\mathbf{A}(\omega)]$, $[\mathbf{B}(\omega)]$ are frequency dependent block diagonal matrix of adequate dimensions, c is a scale factor and $\mathbf{e}_k=[\delta_{ik}]_{1\leq k\leq n}$ is a n dimensional basis vector. Finally, the nonlinear system of differential equations has been converted to a nonlinear system of algebraic equations (9) which provides an error function. The unknown variables $\mathbf{X}(\omega)$ must satisfy $\mathbf{H}(\mathbf{X}(\omega))=\mathbf{0}$, which can be solved using a multidimensional root finding algorithm like Newton-Raphson (Giannini & Leuzzi, 2004; Kundert, 1999; Rizzoli et al. 1992; Suarez & Quéré, 2003). This iterative method, starting from an initial state, iteratively computes the solution by means of a local approximation of the nonlinear function $\mathbf{H}(\cdot)$ to its tangent hyperplane. An outline of the numerical procedure is found in Fig. 4.

For the convergence process to be successful, a good guess of the initial vector $\mathbf{X}_0$ is needed, which can be obtained by the (11) under the assumption of low amplitude driving generators, turning off the nonlinearities giving $\mathbf{X}_0=c\,[\mathbf{A}(\omega)]^{-1} G_s$; $\mathbf{Y}_0$ is then derived by means of Fourier Transform pairs like in (10), and a first estimate for $\mathbf{H}(\mathbf{X})$ is built, which will be subsequently corrected. The routine stops when the norm of the error function is less than a certain threshold, which depends on the prescribed accuracy or when the calculated values for the unknown vector $\mathbf{X}$ at two consecutive steps doesn't differ significantly. In terms of computational resources, the heavier step relies on Jacobian matrix computation (by automatic differentiation) and inversion. Therefore different technique could be used, aimed at solving the linearized sytem resulting from a Newton-Raphson iteration with direct methods or with some advanced techniques (Rizzoli et al., 1997).

Compared to Transient/Time Domain Analysis, Harmonic Balance allows evaluating the steady state response of circuits driven by periodic signals in a faster way. However, the assumption upon which the entire Harmonic Balance mathematical framework holds is that the forced circuit will eventually reach its periodic regime, even though in nonlinear

(symmetrical to zero) positive index ($+k$). Therefore, is possible to set up an Harmonic Balance only for positive frequencies and exploit the last property to compute the coefficients at negative frequencies by a straightforward conjugation, halving the computation time.

circuits very different long term behaviours are possible[13] which possibly coexist. Moreover, if the HB algorithm converges, it won't necessarily do to a stable solution that is observable in reality. On the contrary, should the HB algorithm fail to converge, that wouldn't imply there are no stable solutions. Therefore in order to be sure that the mathematical solution matches the actual one could be necessary to undertake further analysis (Giannini & Leuzzi, 2004).

$$\mathbf{H}\left(\mathbf{X}_h^{(k-1)}(\omega)\right) := [\mathbf{A}(\omega)]\cdot\mathbf{X}_h^{(k-1)}(\omega) - [\mathbf{B}(\omega)]\cdot\mathbf{Y}\left(\mathbf{X}_h^{(k-1)}(\omega)\right) - c\cdot G_s\mathbf{e}_k$$

for $\{-N_m \le h \le N_m\}$ {

evaluate $\|\mathbf{H}(\mathbf{X}_h(\omega))\|$

while $\left(\|\mathbf{H}(\mathbf{X}_h(\omega))\| > \varepsilon\right)$

do {

$$\text{compute } \left[\mathbf{J}_H\left(\mathbf{X}_h^{(k-1)}(\omega)\right)\right] = \frac{\partial\mathbf{H}\left(\mathbf{X}_h^{(k-1)}(\omega)\right)}{\partial\mathbf{X}_h} = [\mathbf{A}(\omega)] - [\mathbf{B}(\omega)]\cdot\frac{\partial\mathbf{Y}\left(\mathbf{X}_h^{(k-1)}(\omega)\right)}{\partial\mathbf{X}_h}$$

$$\text{compute } \mathbf{X}_h^{(k)}(\omega) := \mathbf{X}_h^{(k-1)}(\omega) - \left[\mathbf{J}_H\left(\mathbf{X}_h^{(k-1)}(\omega)\right)\right]^{-1}\mathbf{H}\left(\mathbf{X}_h^{(k-1)}(\omega)\right)$$

$$\text{compute } \mathbf{Y}_h^{(k)}(\omega) := \mathcal{F}\left(\mathbf{y}\left(\mathcal{F}^{-1}\left(\mathbf{X}_h^{(k)}(\omega)\right)\right)\right)$$

$$\mathbf{H}\left(\mathbf{X}_h^{(k)}(\omega)\right) := [\mathbf{A}(\omega)]\cdot\mathbf{X}_h^{(k)}(\omega) - [\mathbf{B}(\omega)]\cdot\mathbf{Y}\left(\mathbf{X}_h^{(k)}(\omega)\right) - c\cdot G_s\mathbf{e}_k$$

evaluate $\|\mathbf{H}(\mathbf{X}_h(\omega))\|$

}

if $\left(\|\mathbf{H}(\mathbf{X}_h(\omega))\| \le \varepsilon\right)$

print the solution $\mathbf{X}_h(\omega)$ }

Fig. 4. Numerical Resolution of the Harmonic Balance system of equations.

3.2 Probe method for oscillator analysis

Harmonic Balance method proves to be very useful when analyzing circuits that are externally forced by time varying RF generators since they provide a first estimate for the circuit solution. However, convergence problems could result from its use, when dealing with autonomous circuits that present self resonant frequencies or sub-harmonic components, like oscillators, since the actual operating frequency and power of the oscillating solution represent two additional unknowns. Motivated by this lack of knowledge HB solutions could be misleading; for example in the analysis of a free running oscillator HB method might converge to a degenerate DC steady state solution, as the only generators left are DC sources. To overcome these difficulties, an idealized component called Auxiliary Generator (AG) which fictitiously plays the role of the missing RF generators, is opportunely inserted in our circuit (Giannini & Leuzzi, 2004; Suarez & Quéré, 2003) to emulate self resonating frequencies or sub-harmonic components, thus forcing the harmonic balance simulator to find the correct solution. It can be represented by an ideal

[13] Sub-harmonic generation, Chaotic behaviour to cite a few, see (Suarez & Quéré, 2003).

sinusoidal voltage source[14] with series impedance (Thevenin equivalent source) that is being connected in parallel to a circuit node N and defined by its amplitude (A_p) frequency (f_p) and phase (φ_p). The series impedance box acts as a very narrowband filter, rejecting all the higher order harmonics while keeping the fundamental[15] f_p. Given the time invariance of the solution waveform in a free-running oscillator the phase of the probe is arbitrarily set to zero, while the amplitude and frequency of the oscillation represent two extra-unknowns that augment the dimension of the HB system of equations (HBE)[16]. Therefore two extra equations have to sought, so that the system (11) is not left underdetermined. Since the admittance of the auxiliary generator Y_p has to be zero when the circuit is operating in the steady state (as if the probe was left disconnected from the oscillator) in order not to perturb its periodic solution, these equations are chosen to be[17] $\mathrm{Re}(Y_p(A_p, f_p))=0$, $\mathrm{Im}(Y_p(A_p, f_p))=0$. With these added equations the HBE returns square, and a solution can be found both for the oscillation amplitude A_0 and fundamental frequency f_0.

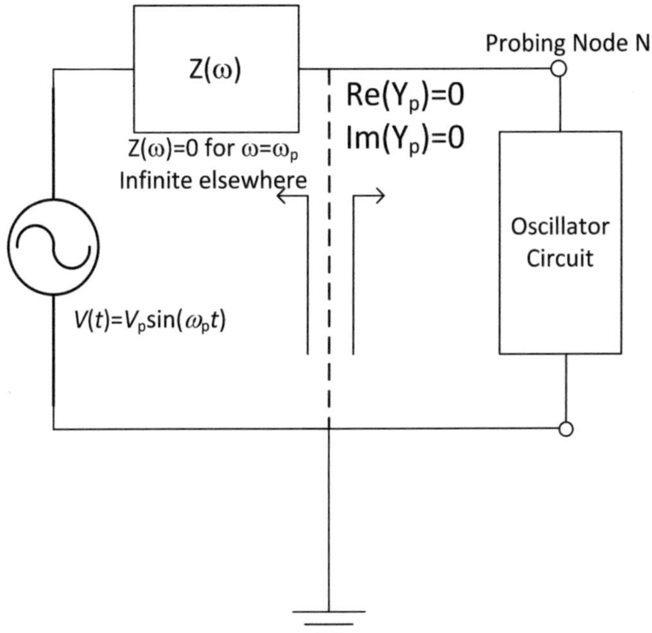

Fig. 5. Auxiliary generator probe.

[14] We could also consider an ideal current source with shunt impedance (Norton equivalent source) that is connected in series to a branch of the circuit to be analyzed. In the following, we will employ only AG having voltage sources.

[15] At frequencies other than f_p the probe is in fact an open circuit.

[16] In the case of the analysis of a synchronized regime instead, the operation frequency is known (*injection frequency*) and the variables to be determined are the phase and amplitude of the probe.

[17] In reality, this condition emerge as a particularization of the famous Kurokawa condition for finding large signal steady state oscillation, which states that $Y_{tot} = Y_{linear} + Y_{nonlinear}=0$ at the oscillation frequency where Y_{linear} is the admittance of the linear subnetwork $Y_{nonlinear}$ is the admittance of the nonlinear sub-network, provided the entire network could be partitioned in that way, and the two sub-networks are connected at a single port. The same condition could be written at the probing port providing, the non-perturbation conditions in terms of Y_p. See (Giannini & Leuzzi, 2004).

In simulation, these constraints are enforced introducing two optimization goals, both for the real and imaginary parts of the admittance. Then the steady state solution is found running the harmonic balance simulation together with the optimization routine in a two tier scheme; the outer tier is constituted by the optimization algorithm, usually a gradient optimizer, that iteratively computes the candidate solutions (A_0, f_0) and pass these values to the inner tier represented by the native HB algorithm, that solve for all the other circuital variables; if at a given step, the distance between the goals and the computed solution is neglectable according to a predefined metric the solution is found, otherwise the search for the optimal point continues, until the maximum number of iterations is reached. In this optimization method great care has to be taken, with respect to the probe insertion point (Brambilla et al., 2010) and to the probe amplitude and frequency initial estimate. Usually the linear analysis frequency estimate works well for achieving convergence of the HB analysis with the probe method (Chang et al., 1991), selecting randomly the initial values for the AG amplitude. On the contrary, probe amplitude is usually guessed in a trial and error scheme, starting with values that are a tenfold less than the biasing voltages, then trying to increase them as long as convergence failure isn't encountered. A more effective scheme to approach the correct amplitude value, consist in assuming that $\mathrm{Im}(Y_p(A, f)) \sim \mathrm{Im}(Y_p(f))$ and that $\mathrm{Re}(Y_p(A, f)) \sim \mathrm{Re}(Y_p(A))$ as for a first order approximation or describing function approach (Giannini and Leuzzi, 2004), in which the susceptance at the node P is *mainly* a function of the probe frequency and the conductance is *mainly* a function of the probe amplitude. In this way, having obtained a frequency estimate from initial linear analysis, this can be kept constant while performing a parametric analysis of the circuit w.r.t the parameter A, plotting the curve $\mathrm{Re}(Y_p(A))$ versus A, and choosing for A_0 the value corresponding to the abscissa intercept that fulfil $\mathrm{Re}(Y_p(A))=0$. Although more complicated, this search method allows one to save many simulation cycles derived by unfruitful attempts.

4. Harmonic balance DVCO analysis

In this section Harmonic Balance simulation in a commercial simulator is combined with the use of an auxiliary probe to analyze the DVCO tuning function. Moreover, numerical continuation techniques, used in conjunction with HB analysis will be employed to show the dependence of the tuning function on some circuit parameters. DVCO nonlinear analysis will thus provide the necessary hints for the synthesis stage.

4.1 DVCO harmonic balance and parametric analysis

The DVCO HB analysis starts with the determination of the discrete resonant frequencies (see section 2) obtained by independently biasing each active device. Therefore having chosen the NE3509M04 the biasing voltage of the active device is chosen in the interval [-0.4 V, 0V] according to its electrical characteristic. Subsequently four HB simulations are performed, choosing the maximum harmonic order $N_{max}=3$, and introducing the auxiliary generator in the vicinity of the DVCO feedback loop to find the oscillation frequencies and output power level. As an acceptable approximation for $\mathrm{Re}(Y_p)$, $\mathrm{Im}(Y_p)=0$ could be considered $-1 \cdot 10^{-18}$ (S) $\leq \mathrm{Re}(Y_p)$, $\mathrm{Im}(Y_p) \leq 1 \cdot 10^{-18}$ (S) which constitute two goal to be fulfilled by the gradient optimizer as detailed in Fig. 6, where the flowchart for finding the DVCO discrete resonant frequencies and the corresponding oscillation power is shown with greater detail.

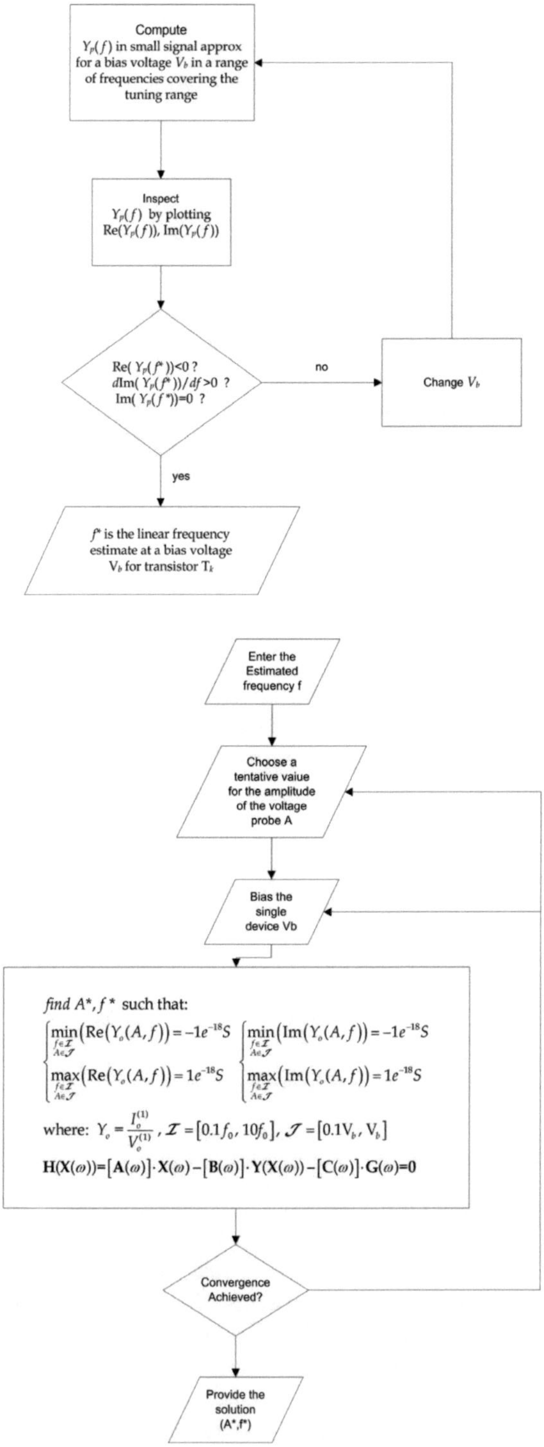

Fig. 6. Double Flowchart for computing the DVCO output spectrum for the discrete oscillation, starting from their small signal frequency estimate.

Active device	Oscillation Frequency	Output power	Biasing Voltage
T_1	2.35023 GHz	2.682 dBm	-0,2015 V
T_2	1.59917 GHz	1.966 dBm	-0.2055 V
T_3	1.13392 GHz	0.762 dBm	-0.1000 V
T_4	0.92538 GHz	-3.017 dBm	-0.1414 V

Table 2. Individual Oscillation Frequencies and Power level.

After the discrete resonant frequencies has been obtained, sensitivity with respect to the lumped components is investigated, by means of a parametric analysis (Collado et al., 2010). To that aim, the inductances (L_g, L_d) and the capacitances (C_g, C_d) are varied in a certain range and the evolution of the oscillatory solutions is traced for each of the four active device separately. Since in our design should be $L_g=L_d=L$, $C_g=C_d=C$, with $1\text{ nH}\leq L\leq 5\text{ nH}$ and $1\text{ pF} \leq C \leq 3\text{ pF}$, the solution is found once the large signal steady state condition $\mathrm{Re}(Y_p(A, f))=0$, $\mathrm{Im}(Y_p(A, f))=0$ and the parameter dependent HBE $\mathbf{H}(\mathbf{X},L,C)=\mathbf{0}$ are simultaneously fulfilled. In a commercial simulator, this has been checked running a parameter swept optimization routine. Since there are two parameters, a double sweep is necessary, the first one for changing the capacitance value and the second (nested in the first) for varying the inductance as illustrated in (Collado et al., 2010).

4.2 DVCO tuning function and stability analysis

Once it has been checked that each of the transistors lead to oscillations distributed along the frequency band, the tuning capabilities of the circuit are analyzed. A modification of the previous routine is used in that the auxiliary generator frequency f_p is increasingly swept in steps (50 MHz is usually enough to ensure convergence) and the gate voltages $V_g^{(i)}, V_g^{(j)}$ ($j=i+1$, i=1,2,3) necessary to synthesize each of the frequencies in the sweep are calculated as additional optimization variables that fulfil the non perturbation conditions $\mathrm{Re}(Y_p(f_p, A_p, V_g^{(i)}, V_g^{(j)}))=0$, $\mathrm{Im}(Y_p(f_p, A_p, V_g^{(i)}, V_g^{(j)}))=0$. In order to obtain the tuning characteristic of the circuit, the frequency tuning band has been divided in three "zones". In each of the zones only two gate voltages are modified to achieve oscillator tuning according to (Divina & Škvor, 1998) while the rest of the transistors are deactivated at $V_g^{(off)}$ = -1V. From an empirical point of view, is noticed that the convergence of the swept optimization process depends, as in the previous case, upon a suitable selection of the initial candidates value for A_p, $V_g^{(i)}$, $V_g^{(j)}$. The tuning voltages of the active sections are initially chosen to be equal to the midpoint of the voltage range in which both devices' transconductance is nonzero, which corresponds to $V_g^{(on)}$ = -0.3V. Subsequently a single frequency point optimization is performed, in order to obtain the values of the oscillation amplitude and frequency for that particular bias configuration. Finally, the frequency swept optimization curves routine starts as described earlier. In case convergence failure should occur, one has to re-initialize A_p, $V_g^{(i)}$, $V_g^{(j)}$ and attempt sweeping the frequency decreasingly. If neither this helps in reaching a solution, numerical continuation techniques have to be invoked.

Frequency Zone	Active Devices	Δ*f* (GHz)
Highest, I	T_1,T_2	1.7 ÷ 2.4
Central, II	T_2,T_3	1.7 ÷ 1.25
Lowest, II	T_3,T_4	1.0 ÷ 1.25

Table 3. Frequency zones and corresponding active devices operating.

The evolution of the gate voltages versus frequency confirms almost perfectly the theoretical predictions (Divina & Škvor, 1998; Škvor et al., 1992), showing a complementary variation on the tuning voltages in pairs. Besides, the DC current consumption reveals a close relationship with the output power level, showing nearly identical trends. It can be observed in (Fig.7) that the output power has larger variations along the edges of each of the frequency tuning zones. A similar behavior can be observed in the DC current consumption. This phenomenon is comprehensible, if it is taken into account that when approaching the border between the zone z-th and (z+1)-th, three gate voltages namely $V_g^{(z-1)}$, $V_g^{(z)}$, $V_g^{(z+1)}$, should come into play in determining the oscillation frequency as it is evident from Table 2. In fact, this abrupt edge-variation has been subsequently corrected, enforcing an additional constraint for the output power and using an additional gate voltage as optimization variable. A heuristic approach was adopted to achieve an output power range that ensured less power fluctuations and no discontinuities in the frequency tuning range. The constraint on the output power has been made tighter in steps. Starting from the initial constraint that limited the output power to fall in the range [-8 dBm, 8 dBm], this range has been halved in the subsequent steps; having checked the frequency tuning curves to guarantee the circuit did not lose its tuning capabilities. The result of applying this optimization process can be seen in Fig.8. The necessary gate voltages in order to tune the frequency and at the same time maintain the output power characteristic inside the variation limits are represented. The output power variation along the frequency tuning band is shown. After numerous simulations, the output power goal range has been chosen to be [3 dBm, 7 dBm], assuring a maximum difference in the output power of 4 dB for nearly 96% of the tuning bandwidth.

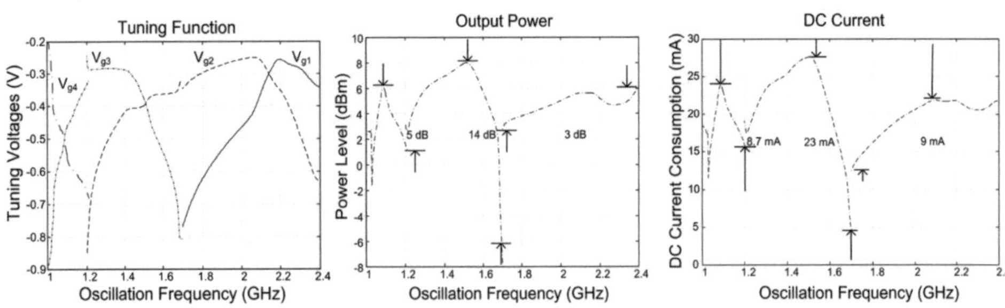

Fig. 7. Tuning Function, Output Power level, DC current for the simulated DVCO.

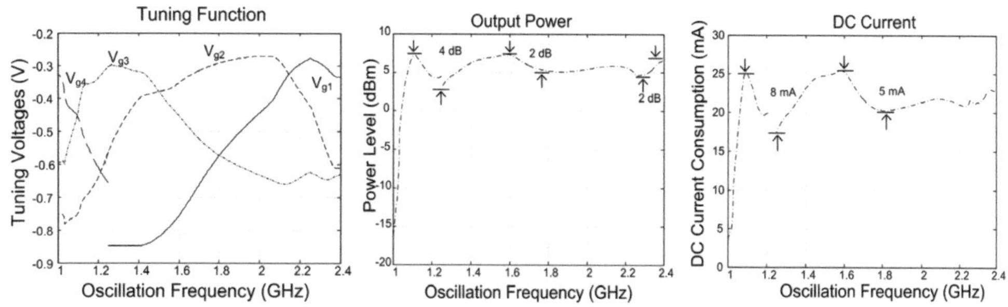

Fig. 8. Tuning range obtained varying the gate voltages, after power optimization.

A power overshoot is still present at the very beginning of the band, between 1 and 1.1 GHz while intermediate power drops have been completely removed. Furthermore in more than 60% of the tuning band (from 1.35 GHz to 2.4 GHz) power has shown limited fluctuations: 2 dB in the frequency span from 1.35 GHz (approx.) to 2.25 GHz (approx.) and the same amount in the interval between 2.25 and 2.4 GHz, with an absolute variation of 2.5 dB. Similar remarks are valid for the DC current consumption. A current overshoot takes place at the beginning of the band, while the maximum absolute variation in the dissipated current is 8 mA and occurs locally (at 1.25 GHz approx). In the rest of the band variations are kept limited to 5 mA.

The assumption made up to now is that if the convergence of the harmonic balance system of equations and the condition at the probe port $\mathrm{Re}(Y_p)=0$, $\mathrm{Im}(Y_p)=0$ are simultaneously satisfied the solution will be unique. However, due to the particular configuration of the DVCO, multiple oscillating modes are possible, each one characterized by different power level (Acampora et al., 2010). Thus, the solution represented in fig. 9-10 represents in effect one member of a family of solutions whose stability needs to be investigated. In this regard, double parametric sweeps have been performed (Collado et al., 2010) choosing as parameters ($V_g^{(1)}$, $V_g^{(2)}$) in the first zone ($V_g^{(2)}$, $V_g^{(3)}$) in the second zone and ($V_g^{(3)}$, $V_g^{(4)}$) in the third zone and iteratively running HB simulations for $-1\mathrm{V} \leq (V_g^{(i)}, V_g^{(i+1)}) \leq 0\mathrm{V}$. The curves shown in (Collado et al., 2010) illustrate that it's possible to synthesize the same frequency with more than one bias combination.

5. DVCO implementation and measurements results

After the nonlinear simulations were carried out, a DVCO prototype was fabricated taking advantage of local facilities (Acampora et al., 2010; Collado et al., 2010). The layout of the device was realized on a 20 mil (0.5 mm) thick Arlon A25N substrate, where the microstrip lines' geometries were grooved using a mechanical milling machine. Lumped inductors and capacitors from TOKO were soldered on the circuit, which was tested several times in order to ensure it would comply with the expectations. For the measurement phase, five independent digitally controlled voltage sources were used; four of them were the negative tuning voltages to adjust the oscillation frequency, and the last one was the constant drain line voltage, which was common to all the sections. The circuit is shown in Figure 9.

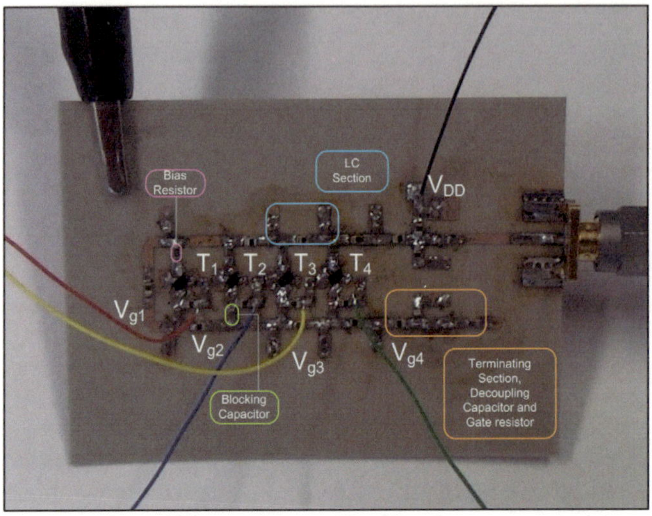

Fig. 9. A prototype of the DVCO.

For keeping track of the oscillation frequency and power an Agilent Spectrum Analyser E4448A (Fig. 10) was used, having an operating bandwidth from 3 GHz to 50 GHz. The results are shown in the pictures below (Fig. 11). It is seen that experimental tuning characteristic, power and DC current consumption curves match the theoretical predictions, although the tuning bandwidth is somehow shifted towards lower values. This effect is most likely due to a mismatch between vendor's model and real characteristic for the discrete component and active devices. Moreover, in the schematic used for simulation, coupling effect between adjacent lines has been neglected, and since resistors model from the vendor were unavailable, they have been approximated with ideal elements.

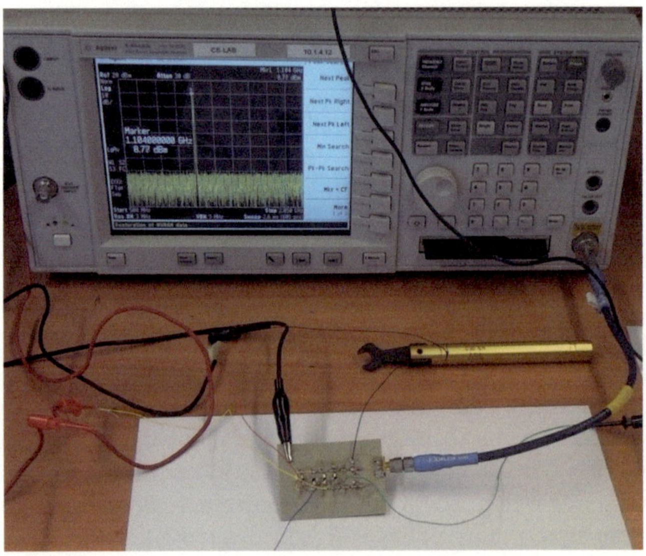

Fig. 10. Measurement benchmark for deriving the tuning curves, output power and DC current for the DVCO.

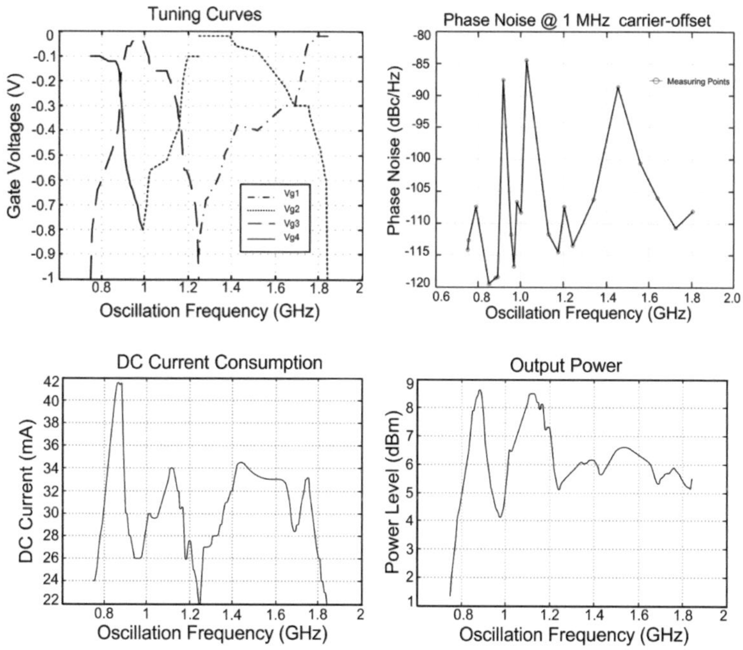

Fig. 11. Measurement results in terms of tuning curves, output power, DC current consumption at phase noise at 1MHz offset from the carrier, across the tuning band.

In the first prototypes, much attention was devoted to kill some parasitic oscillations that prevented the DVCO from having a continuous tuning function. In these cases, many simulations were made to find out the optimal value of the drain bias resistor that would eliminate the spurious oscillation.

However, still the experimental tuning function suffered from instabilities and hysteresis phenomena. Ambiguous tuning function behaviour has been experimentally substantiated. In fact keeping one voltage fixed at a certain value and changing a neighbouring one, the output oscillation didn't change its frequency but its power. In some cases, oscillations "jumps" in frequency have been noticed. It has been observed empirically that all these issue might be mitigated at least, if the drain voltage V_{DD} is varied in order to keep the DC current I_{DD} constant and to low values (on the order of 10 mA). Maintaining I_{DD} low, appears to be beneficial also in terms of phase noise, where is seen that poorest phase noise performance corresponds to those oscillations whose DC current consumption is elevated.

6. Conclusion

The need for flexible/smart protocols in which radio resources are dynamically allocated among users across a wide frequency band, calls for improvement on communication subsystems employed on user equipments. Since oscillators, VCOs and other oscillator-driven systems are remarkably important elements in every RF front-end; it is believed that they should be endowed with wideband spectrum sensing capability to accommodate the needs of Cognitive Radio technology.

To that aim, Distributed Voltage Controlled Oscillators have been investigated in this chapter, pointing out their general features and discussing with greater detail one circuit that provides very wideband tuning capabilities namely the "reverse mode DVCO". Starting from consolidated results, our purpose has been to shed new light on DVCO operation with the help of nonlinear analysis and simulation tools, which has allowed to confirm the theoretical predictions already established and to proceed further in investigating the tuning function, the output power variation across the tuning band, and DC current consumption. Moreover, a prototype of the DVCO has been fabricated utilizing discrete components, on a microstrips printed circuit board. The measurements results matched adequately the nonlinear simulations. Nevertheless more research will be needed to discover a suitable tuning algorithm.

Future research tasks include the implementation of a Distributed Divider, and studying the injection locking properties of the DVCO (Paciorek, 1965; Tsang & Yuen, 2004). Early tests with a slightly modified version of the DVCO shown in this chapter manifested its super harmonic and sub harmonic injection locking properties, allowing the synchronization of the DVCO output with an external signal whose frequency is a multiple (double or triple) with respect to the fundamental in the case of super-harmonic injection locking or a sub-multiple, (half of the fundamental) in the case of sub-harmonic injection locking.

7. References

Acampora, A.; Collado, A. & Georgiadis, A.(2010). Nonlinear analysis and optimization of a Distributed Voltage Controlled Oscillator for Cognitive Radio, *Proceedings of the 2010 IEEE International Microwave Workshop Series (IMWS) on RF Front-ends for Software Defined and Cognitive Radio Solutions*, vol., no., February 2010, pp.1-4

Akyildiz, I. F.; Lee, W.; Vuran, M. C. & Mohanty, S. (2006). NeXt generation/dynamic spectrum access/cognitive radio wireless networks: a survey. *Elsevier Journal on Computer Networks*, vol. 50, no. 13, May 2006, pp.2127 – 2159

Brachtendorf, H.G.; Welsch, G. & Laur, R. (1998). A time-frequency algorithm for the simulation of the initial transient response of oscillators. *Proceedings of the 1998 IEEE International Symposium on Circuits and Systems, (ISCAS '98)* vol.6, no., 31 May-3 Jun 1998, pp.236-239

Brambilla, A.; Gruosso, G.; Gajani, G.S. (2010). Robust Harmonic-Probe Method for the Simulation of Oscillators, *IEEE Transactions on Circuits and Systems I: Regular Papers*, vol.57, no.9, Sept. 2010, pp.2531-2541

Chang, C.-R.; Steer, M.B.; Martin, S. & Reese, E. Jr. (1991). Computer-aided analysis of free-running microwave oscillators, *IEEE Transactions on Microwave Theory and Techniques*, vol. 39, no. 10, October 1991 pp.1735-1745

Christoffersen, C.E.; Ozkar, M.; Steer, M.B.; Case, M.G. & Rodwell, M. (1999). State-variable-based transient analysis using convolution. *IEEE Transactions on Microwave Theory and Techniques*, vol.47, no.6, pp.882-889, June 1999

Cohen, L.D. (1979). Varactor Tuned Gunn Oscillators with Wide Tuning Range for the 25 to 75 GHz Frequency Band. *Proceedings of the 1979 IEEE MTT-S International Microwave Symposium Digest*, , vol., no., April 30-May 2, 1979, pp.177-179

Collado, A.; Acampora, A. & Georgiadis, A. (2010). Nonlinear analysis and synthesis of distributed voltage controlled oscillators. *International Journal of Microwave and Wireless Technologies*, vol.2, no.2, April 2010, pp.159-163

Divina L. & Škvor Z. (1995). Experimental verification of a distributed amplifier oscillator, *Proceedings of the 25th European Microwave Conference (EuMC) 1995*, Kent, U.K. Nexus Media Limited, September 1995, pp. 1163–1167

Divina, L. & Škvor, Z. (1998). The Distributed Oscillator at 4 GHz. *IEEE Transactions On Microwave Theory And Techniques*, vol. 46, no. 12, December 1998, pp 2240-2243

Ferrus, R.; Sallent, O. & Agusti, R. (2010). Interworking in heterogeneous wireless networks: Comprehensive framework and future trends, *IEEE Wireless Communications Magazine*, vol.17, no.2, April 2010, pp.22-31

Frei, S.; Fuhrmann, W.; Rinkel, A. & Ghita, B.V. (2011). Improvements to Inter System Handover in the EPC Environment, *Proceedings of the 2011 4th IFIP International Conference on New Technologies, Mobility and Security (NTMS)*, vol., no., February 2011, pp.1-5, 7-10

Giannini F. & Leuzzi G. (June 28, 2004). *Nonlinear Microwave Circuit Design*, Wiley, ISBN: 978-0470847015

Ginzton E. L.; Hewlett, W. R.; Jasberg, J. H. & Noe, J. D. (1948). Distributed Amplification *Proceedings of the IRE*, August 1948, pp. 956-69

Heydary, P. (2005). Design and Analysis of A Performance-Optimized CMOS UWB Distributed LNA. *IEEE Journal Of Solid-State Circuits*, vol.42, no.9, September 2007, pp.1892-1905

Kundert K.S. (1999). Introduction to RF simulation and its application, *IEEE Journal on Solid-State Circuit*, 34, September 1999, pp. 1298–1319, 14.09.2011, Available also from: http://www.designers-guide.org/Analysis/rf-sim.pdf (October 2010)

Kundert, K.S. (1997). Simulation methods for RF integrated circuits, *Computer-Aided Design, 1997. Digest of Technical Papers., 1997 IEEE/ACM International Conference on* , vol., no., November 1997, pp.752-765, 9-13

Kundert, K.S.; White, K.S. & Sangiovanni-Vincentelli, A. (March 31, 1990). *Steady-state methods for simulating analog and microwave circuits*, Kluwer Academic Publishers, March 1990, Springer reprint of hardcover 1st edition, December 2010, ISBN: 978-1441951212

Ngoya, E. & Larcheveque, R. (1996). Envelope transient analysis: A new method for the transient and steady state analysis of microwave communication circuits and systems, *Proceedings of the IEEE MTT-S International Microwave Symposium Digest*, 1996, vol.3, no., June 1996, pp.1365-1368

Ngoya, E.; Suarez A.; Sommet R. & Quéré, R. (1995). Steady state analysis of free or forced oscillators by harmonic balance and stability investigation of periodic and quasi-periodic regimes. *Wiley International Journal on Microwave and Millimeter Wave Computer Aided Engineering* ,vol. 5, no. 3, May 1995, pp. 210–223

Paciorek, L.J. (1965). Injection locking of oscillators, *Proceedings of the IEEE* , vol.53, no.11, November 1965, pp. 1723- 1727

Percival, W. S. (1937). Thermionic Valve Circuits. *British Patent Specification no. 460,562*, filed 24 July 1936, granted January 1937

Peterson, D.F. (1980). Varactor Properties for Wide-Band Linear-Tuning Microwave VCO's. *IEEE Transactions on Microwave Theory and Techniques*, vol.28, no.2, Februay 1980, pp. 110- 119

Pozar D. (February 5, 2004). *Microwave Engineering*, Wiley, New York, ISBN: 978-0471448785

Rizzoli, V. & Neri, A. (1988). State of the art and present trends in nonlinear microwave CAD techniques *IEEE Transactions on Microwave Theory and Techniques*, vol. 36, February 1988, pp. 343–356

Rizzoli, V.; Lipparini, A.; Constanzo, A.; Mastri, F.; Neri, A. & Massoti, S. (1992). State of the art harmonic balance simulation of forced nonlinear microwave circuits by the piecewise technique, *IEEE Transactions on Microwave Theory and Techniques*, vol. 40, no., January 1992, pp. 12–28

Rizzoli, V.; Mastri, F.; Cecchetti, C.; Sgallari, F. (1997). Fast and robust inexact Newton approach to the harmonic-balance analysis of nonlinear microwave circuits, *IEEE Microwave and Guided Wave Letters,* vol.7, no.10, October 1997, pp.359-361

Safarian, A. Q.; Yazdi, A. & Yedari, P. (2005). Design and Analysis of an Ultrawide-Band Distributed CMOS Mixer. *IEEE Transactions On Very Large Scale of integration (VLSI) Systems,* vol.13, no.5, May 2005, pp 618-629

Silverberg, M. & Wing, O. (1968). Time Domain Computer Solutions for Networks Containing Lumped Nonlinear Elements. *IEEE Transactions on Circuit Theory,* vol.15, no.3, September 1968 pp. 292- 294

Škvor, Z.; Saunders, S.R. & Aitchison, C.S. (1992). Novel decade electronically tunable microwave oscillator based on the distributed amplifier. *IEEE Electronics Letters,* vol.28, no.17, August 1992, pp.1647-1648

Sobhy, M. I. & Jastrzebski, A. K. (1985). Direct Integration Methods of Non-Linear Microwave Circuits. *Proceedings of the 15th European Microwave Conference, 1985,* vol., no., September 1985, pp.1110-1118

Suarez, A. & Quéré, R. (January 2003). *Stability Analysis of Nonlinear Microwave Circuits,* Artech House publishers, ISBN: 978-1580533034

Sun, Y. (1972). Generation of Sinusoidal Voltage (Current)-Controlled Oscillators for Integrated Circuits. *IEEE Transactions on Circuit Theory,* vol.19, no.2, March 1972, pp. 137- 141

Tsang, K.F. & Yuen, C. M. (2004). A 2.7 V, 5.2 GHz frequency synthesizer with 1/2 -subharmonically injection-locked distributed voltage controlled oscillator. *IEEE Transactions on Consumer Electronics,* vol.50, no.4, November 2004, pp. 1237- 1243

Wireless World Research Forum Working Group 6 (2004). Cognitive Radio, Spectrum and Radio Resource Management (White Paper), 14.09.2011, Available from http://www.wireless-world-research.org/fileadmin/sites/default/files/about_the_forum/WG/WG6/White%20Paper/WG6_WP4.pdf

Wong, T.T.Y. (1993). *Fundamentals of Distributed Amplification,* 1993, Artech House, London, ISBN: 978-089-0066-15-7

Wu, H. & Hajimiri, A. (2000). A 10 GHz CMOS distributed voltage controlled oscillator, *Proceedings of the IEEE 2000 Custom Integrated Circuits Conference (CICC).,* vol., no., May 2000, pp.581-584

Wu, H. & Hajimiri, A. (2001). Silicon-Based Distributed Voltage-Controlled Oscillators. *IEEE Journal Of Solid-State Circuits,* vol. 36, no. 3, March 2001, pp 493-402

Yuen, C. M. & Tsang, K. F. (2004). A 1.8-V distributed voltage-controlled oscillator module for 5.8-GHz ISM band. *IEEE Microwave and Wireless Components Letters,* vol.14, no.11, November 2004, pp. 525- 527

Zhou, B.; Rhee, W. & Wang, Z. (2011). Relaxation oscillator with quadrature triangular and square waveform generation, *IEEE Electronics Letters* , vol.47, no.13, June 2011, pp.779-780

2

Control Plane for Spectrum Access and Mobility in Cognitive Radio Networks with Heterogeneous Frequency Devices

Nicolás Bolívar and José L. Marzo
Universitat de Girona
Spain

1. Introduction

One of the main problems that were identified for the insertion of future wireless applications is that an apparent scarcity exists in the wireless frequency spectrum. However, studies demonstrated that the spectrum is inefficiently distributed as opposed as scarce (Shukla et al, 2007). In Fig. 1, the difference between spectrum scarcity and spectrum misuse is shown. In the first scenario, a new application, represented by U6, wants to use the wireless spectrum but has no space to communicate. In the second scenario, the same application is not able to communicate due to an inefficient distribution.

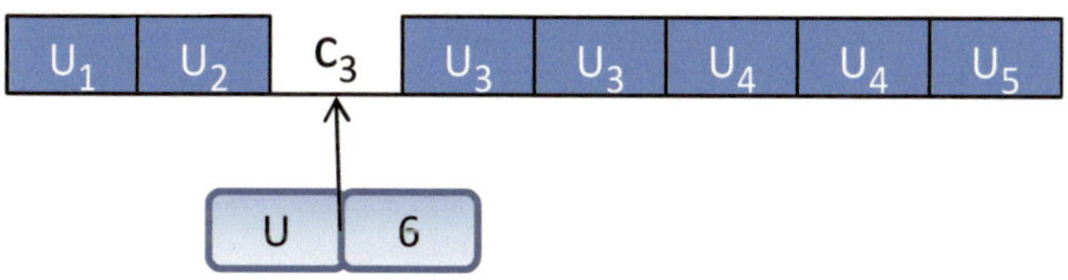

Fig. 1. a) Spectrum Scarcity

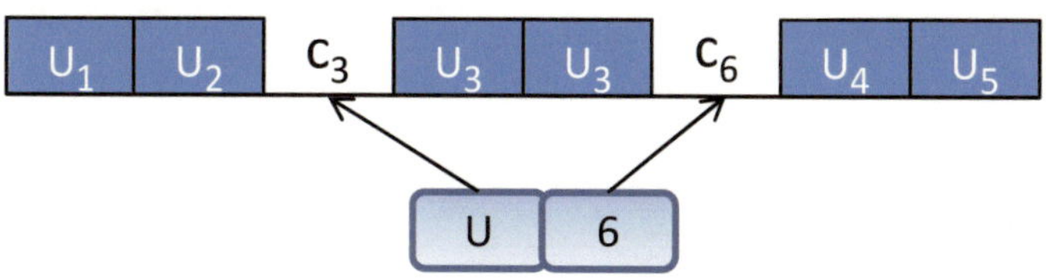

Fig. 1. b) Spectrum misuse

Cognitive Radio Networks (CRN) were defined as networks where user devices are able to adapt to the environment (Mitola & Maguire, 1999). Among the adaptability characteristics, CRN should use the spectrum in an opportunistic manner. In order to do so, Cognitive Radio (CR) devices should be able to recognize spectrum holes, and to use Dynamic Spectrum Access (DSA) capabilities through those frequency slots. Therefore, the use of CRN is an excellent candidate for solving the apparent scarcity problem.

In general, a CRN should be able to perform 4 tasks efficiently: spectrum sensing, spectrum decision, spectrum sharing, and spectrum mobility. Spectrum sensing refers to the identification of the most likely white spaces or spectrum holes in a specific moment. Spectrum decision refers to the process of deciding in which holes to allocate communications (Akyildiz et al, 2008). The spectrum sharing function consists on maximizing the Cognitive Radio Users (CRUs) performance without disturbing Primary Users (PUs) and other CRUs (Akyildiz et al, 2008; Wang et al, 2008). In our work, we consider the spectrum decision and spectrum sharing as parts of an entity called spectrum access. Spectrum mobility is the CRU ability to leave a frequency portion of the spectrum occupied when a PU starts using the same part of the spectrum and then, to find another suitable frequency hole for communication (Akyildiz et al, 2008).

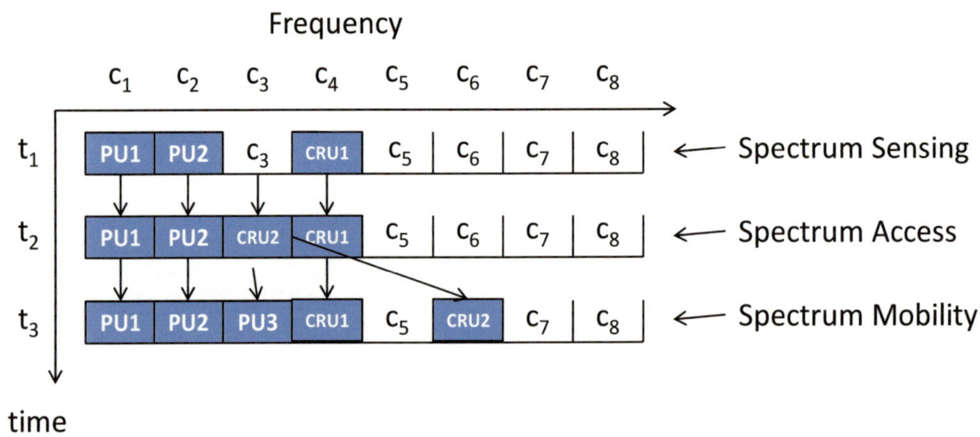

Fig. 2. Spectrum Functions

2. Control plane

In order to efficiently distribute the CRUs in their corresponding channels without interfering both previous CRU communications and PU in their licensed bands, coordination and control signals must be continuously sent in the CRN. The need of a control plane has been discussed in (Jing & Raychaudhuri, 2007). However, to the authors' best knowledge, there is not a review in the literature about the alternatives for transmitting control messages. The closest ones are presented in (Chowdury & Akyildiz, 2011) and in (Theis et al, 2011) for the rendezvous problem, i.e. user discovery in a DSA environment. In this chapter, we provide a quick review about the control plane alternatives combining the classifications defined by (Chowdury & Akyildiz, 2011; Theis et al, 2011) and expanding them to consider all the control plane alternatives.

2.1 Classification

There have been different approaches for transmitting control signals for CRN. Since a dedicated common control channel might not be available at all times, several techniques have been discussed for the 'control channel' problem. However, control signals are basically transmitted through the following strategies.

According to the specialization of the channel, we can divide the control messaging strategies in dedicated and shared control messaging; according to the number of channels used for control messaging, in single (common) and multiple control messaging. According to the frequency-changing nature of the channels, in fixed and hoping control messaging. Finally, according to the lever of power, we can divide them in underlay and overlay control messaging.

The utilization of dedicated control messaging implies the presence of specialized control channels, while the shared control messaging indicates that the same channels are used for both control and communication messages. In single, or common, control messaging only one channel is used for transmitting control messages. On the other hand, multiple control messaging implies that at least two channels are used at the same time for control message transmission. Fixed control messaging indicates that the channel(s) for the transmission of control messages are the same for the whole period of time. Hoping control messaging is presented when the channels used for control messaging vary over time. Finally, underlay control messaging indicates that the control messages are sent below a power threshold, while overlay control messaging indicates that these messages are sent only through available channels. In this section, these classes of messaging are explained in detail.

2.2 Dedicated Control Messaging (DCM)

This approach is the equivalent of having Dedicated Control Channels (DCCs). In this case, the control messages are transmitted separately from the data messages, i.e. through different channels. In Fig. 3, an example of the dedicated DCM with one DCC is shown.

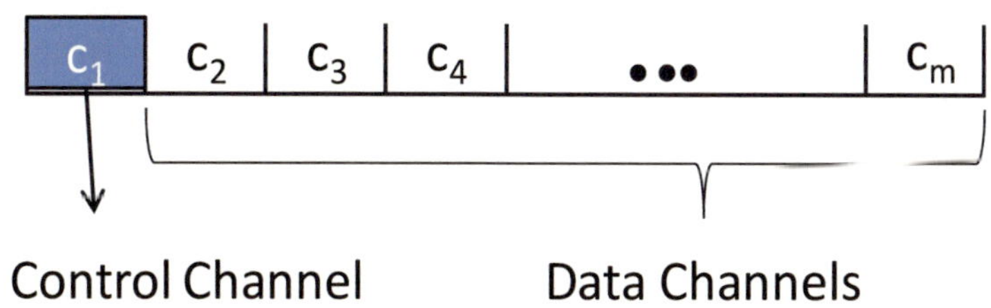

Fig. 3. Dedicated control messaging

The advantage of using DCM is that no additional processing is needed to differentiate the control messages from the data ones. The main disadvantage is that in the case that control messaging is not needed at every time slot, a waste of resources, which is a critical issue for CR as a solution of the wireless spectrum scarcity problem, is present.

2.3 Shared Control Messaging (SCM)

On the other hand, in the SCM the same channels are used for transmitting both control and data messages. Different strategies must be taken into account for separating both types of transmission. In Fig. 4, an example of a frequency-division for the control transmission in the same data channels is shown. Other strategies include time-division and code-division, among others.

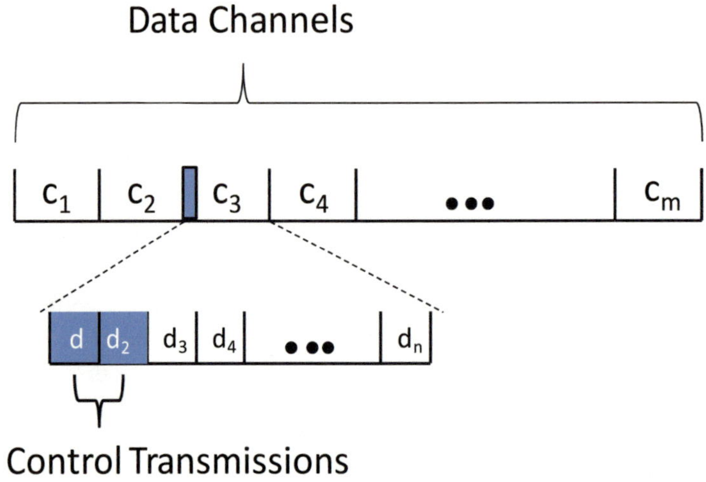

Fig. 4. Shared control messaging (Frequency-division)

In the case from Fig. 4, two sub-slots are used for transmitting control messages. In this scenario, the resources might be used more efficiently but more complex processing is needed, compared to DCM.

2.4 Single (Common) Control Messaging (CCM)

In this case, only one channel is used for transmitting control messages. To be a suitable alternative for transmitting control messages, CCM requires that all devices must have at least one available channel in common for being the Common Control Channel (CCC). In Fig.5, c_3 is selected among all the data channels for transmitting the control messages as a CCC.

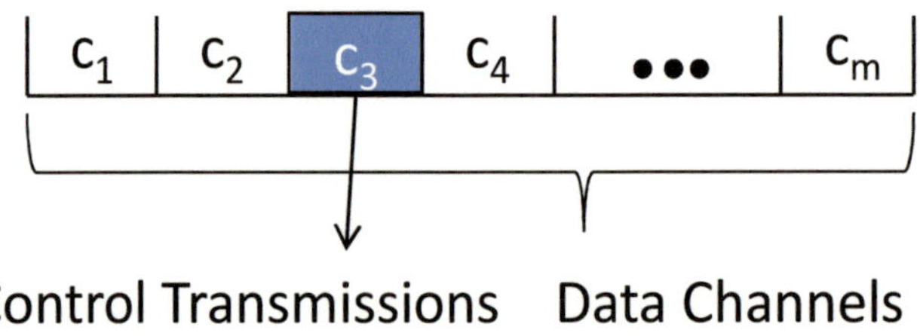

Fig. 5. Common control messaging

The main problems that might arise for this strategy in CRN are that the control channel could be also affected by the presence of PU. For heterogeneous devices, this approach might not be useful since the devices in the CRN could present different sets of channels.

2.5 Multiple Control Messaging (MCM)

In this case, multiple channels are used for transmitting control information. This approach is very useful when not all of the users share the same characteristics such as frequency bands and location. In Fig. 6, c_1 and c_3 are the channels selected for control transmissions.

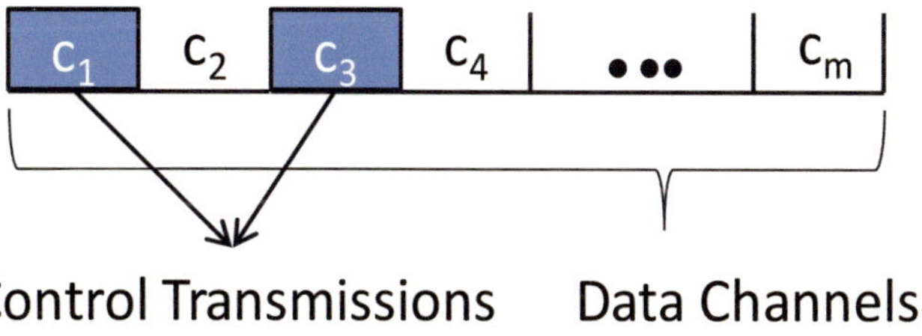

Fig. 6. Multiple control messaging

The main disadvantage of MCM is that the users must be able to receive control messages in different channels. A special case of the MCM is the clustered approach, in which users are divided into clusters according to a specified characteristic. In Fig. 7, an example of the clustered control messaging is shown.

2.5.1 Clustered approach

Let us suppose a centralized CRN covering 8 CRUs: U1, U2, ... , U8, each of them using different sets of frequency channels. A Central Cognitive Base Station (CCBS), in this case, BS, should assign them the necessary channels to transmit control information.

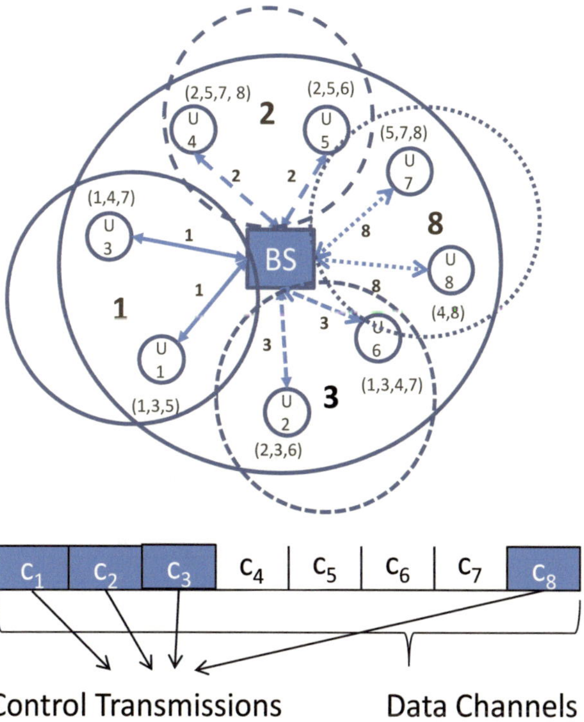

Fig. 7. Clustered control messaging

In the example shown in Fig. 7, four channels are selected for transmitting control information. Channel 1 is used for U1 and U3, channel 2, for U4 and U5. Channel 3, for U2 and U6, and channel 8, for U7 and U8.

2.6 Fixed Control Messaging (FCM)

In this scenario, the same sets of channels are used to transmit control messaging over time. The advantage of FCM is that the receivers are set in the same frequencies. In Fig. 8, c_3 is chosen to be the channel used for control transmissions.

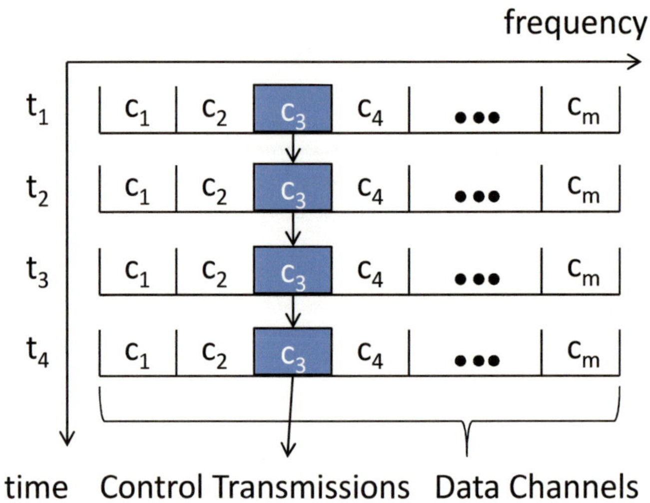

Fig. 8. Fixed control messaging

The main disadvantage of the FCM is that the channels used for control might be also affected by the presence of PU and could be unavailable for control transmission in critical moments.

2.7 Hoping Control Messaging (HCM)

In this scenario, the users change along time the channels they use to receive control messages. In Fig. 9, a sequence for choosing the channel used for control messaging is shown.

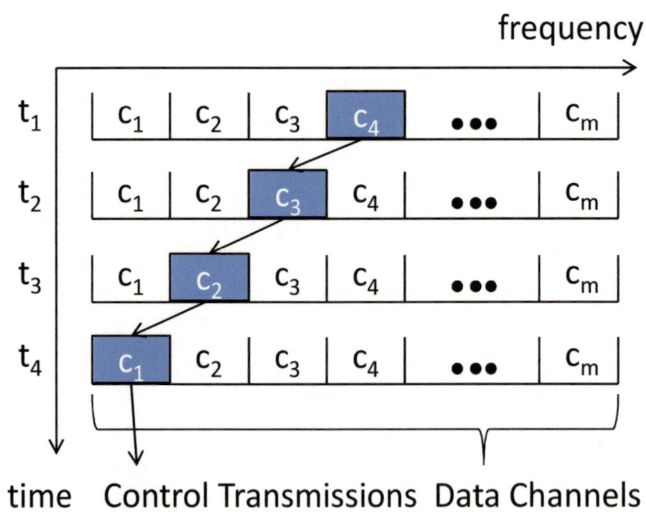

Fig. 9. Hoping control messaging

The main advantage of the HCM is that if a PU is present in a channel that was assigned for control transmissions, another channel might be selected for control messaging. The main disadvantages are that both extra information and a synchronization mechanism are needed.

2.7.1 Default Hoping (DH-HCM)

In this hoping mechanism, a pattern for the control channel is introduced. CRUs should be aware of the sequence beforehand. In Fig. 10, besides the frequency vs. time representation, the time vs. frequency representation is shown, to represent continuity in time.

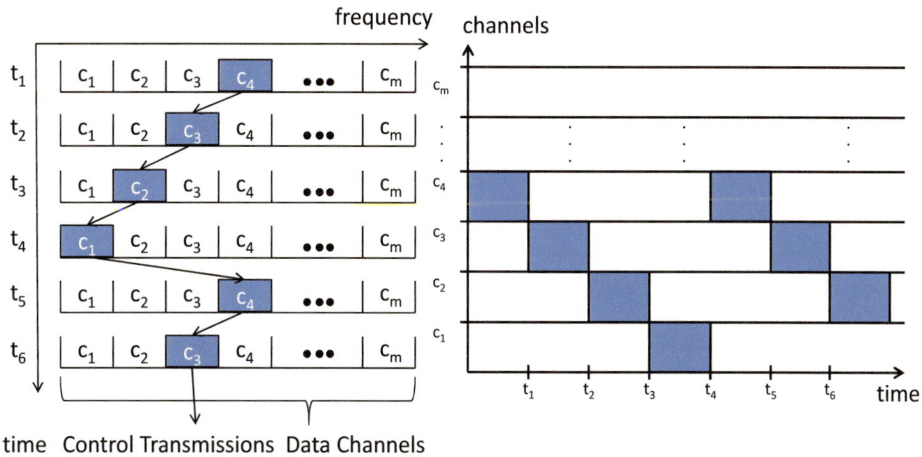

Fig. 10. Default Hoping

2.7.2 Common Hoping (CH-HCM)

In this hoping mechanism, two or more users, after negotiating, hop to the same channel in order to share control information. In this scenario, the next channel(s) used for control information is chosen from the set of available ones. In Fig. 11, both representations in frequency vs. time and vice versa are presented.

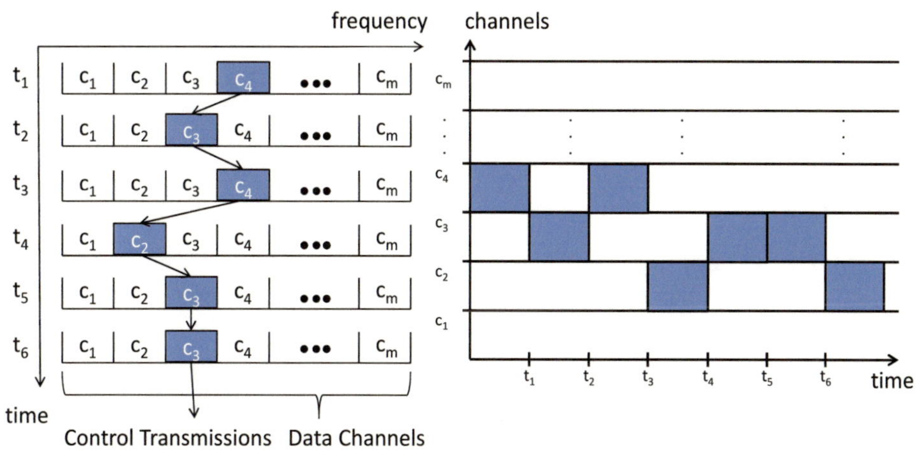

Fig. 11. Common Hoping

2.8 Underlay Control Messaging (UCM)

This approach is the equivalent of transmitting control signals below a power threshold among one or more channels. An example of the UCM is shown in Fig. 12.

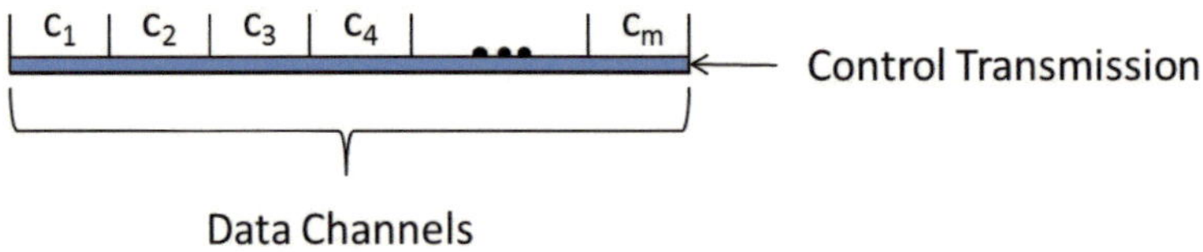

Fig. 12. Underlay control channel

In this case, if a PU requests to use its licensed channel, the control signals should not interfere with the PU transmission. The main advantage is that control transmissions should be performed at any time. The main disadvantage is that the power limit should be chosen carefully in order to guarantee that no licensed user is disturbed.

2.9 Overlay Control Messaging (OCM)

This approach is the equivalent of using Opportunistic Spectrum Access (OSA), i.e. a channel could be used for transmitting control information only if in that channel power indicates that the channel is unoccupied, or DCCs. An example of an OCM using OSA is shown in Fig. 13.

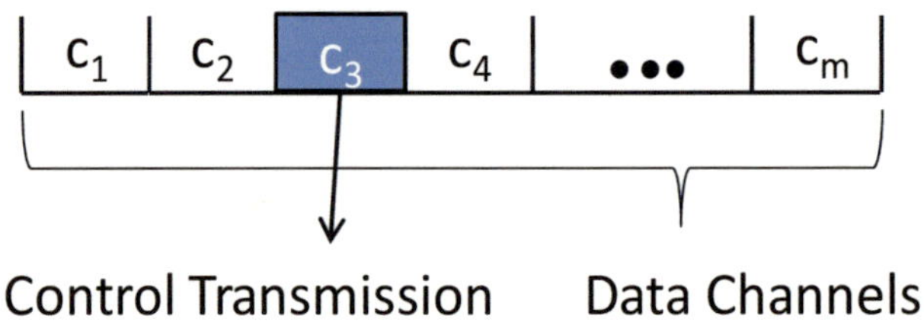

Fig. 13. Overlay control channel

The main problem that might arise for this strategy is that in the case of a DCC, resources might be wasted. On the other hand, in the OSA case, a power level might be misinterpreted in the sensing part and cause interference, and in presence of PU, a hoping mechanism might be needed to be activated to avoid the interference.

2.10 Discussion

In general, each strategy for control messaging is classified into four of the previous categories. For example, when only one channel is used for transmitting control information all the time, and in this channel no data is sent, this approach can be classified into DCM, CCM, FCM and OCM.

Another example is transmitting control information below a threshold in a fixed set of channels that are also used in an overlay manner for CR. In that case, the control approach can be classified as SCM, MCM, FCM and UCM.

Keep in mind that some of the strategies, while not apparent, might solve problems that arise in different circumstances. For example, a common problem for cognitive radio ad-hoc networks (CRAHNs) is the discovery of the channel when HCM is selected due to PU presence. In the case, DH-HCM can be an excellent strategy considering that although time synchronization among the CRUs is needed, the discovery of the channel where control messages are sent is solved because the CRUs could know where to 'listen' for control information at any specific moment. The difference between the Centralized CRN approach and the CRAHNs can be seen if Fig. 14.

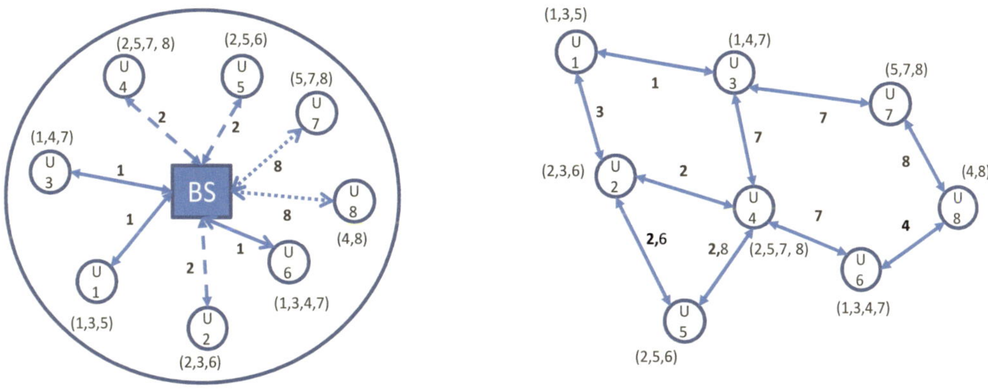

Fig. 14. Centralized and Ad-Hoc CRNs.

In the next section, a model proposed to transmit control information to heterogeneous users in a centralized CRN while using OSA is presented. This model uses SCM, MCM, HCM and OCM.

3. Model

3.1 Antecedents

There have been different approaches for transmitting control and coordination signals for spectrum access and mobility in CRN. Since a dedicated common control channel might not be available at all times, several techniques have been discussed for the *control channel* problem. For a CRN, the relationship between the spectrum functions might be represented as in Fig. 15.

The utilization of beacons was suggested as a solution for spectrum access by using these beacons to control the medium access of the network devices into the frequency bands (Hulbert, 2005). Architectures with more than one beacon have been proposed to improve performance (Mangold et al, 2006). In these proposals, the beacons are sent by the PU through a cooperative control channel or a beacon channel, with the latter being considered a better option in (Ghasemi & Sousa, 2008). This approach has two main disadvantages for implementation in a CRN with today's available technologies; the first is that a new set of primary users must exist or new hardware must be developed since the PUs should inform the nearby CRU about their presence, and the second disadvantage is that a new channel must be reserved for the beacon signals. In Fig. 16, a division in channels and sub-channels is presented in order to use some of the sub-channels for beacon transmission.

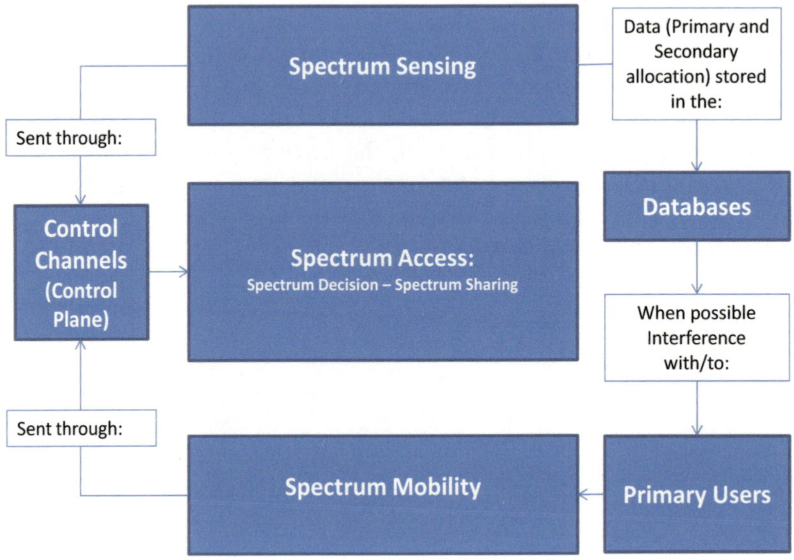

Fig. 15. Spectrum Functions and Control Plane

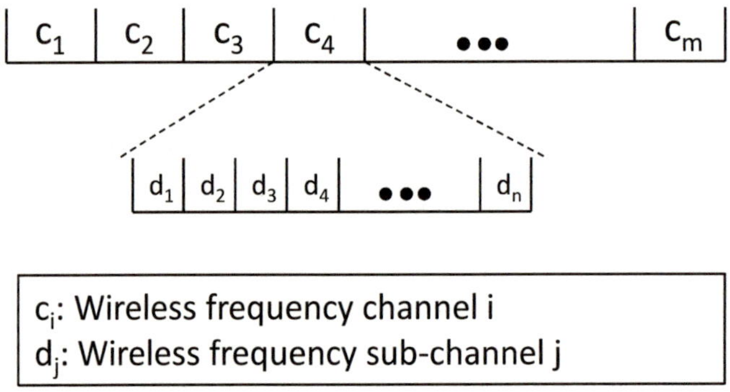

Fig. 16. a)Wireless Frequency Channel-Sub channel Division

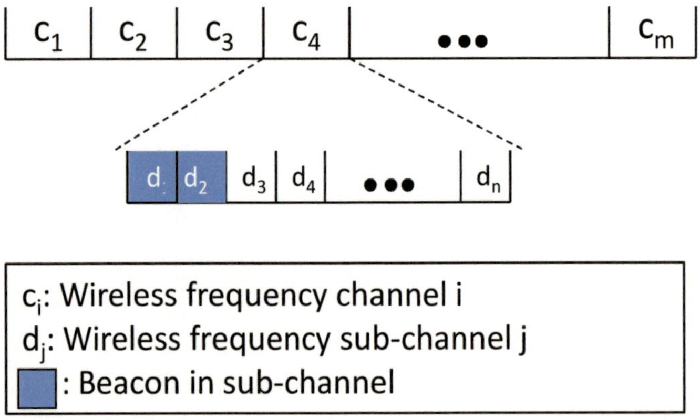

Fig. 16. b)Beacons in Wireless frequency sub-channels

A Cognitive Pilot Channel (CPC) is a solution proposed in the E2R project for enabling communication among heterogeneous wireless networks (Bourse, 2007). The CPC consists on controlling frequency bands in a single or various "pilot" channels, which is analogue to the beacon proposal. In both CPC and beacons proposal, there are "in-band" transmission, i.e. information transmitted in the same logical channels of the data transmission, and "out-band" transmission, i.e. information transmitted in different channels of the data transmission (Sallent et al, 2009). Studies have been conducted to define the quantity of information that should be transmitted in the CPC, the bandwidth for each CPC, and the "out-band" and the "in-band" transmission or other solutions with a combination of both (Filo et al, 2009; Pérez-Romero et al, 2007; Sallent et al, 2009). In Fig. 17, we can see the difference of the in-band and out-band control transmission.

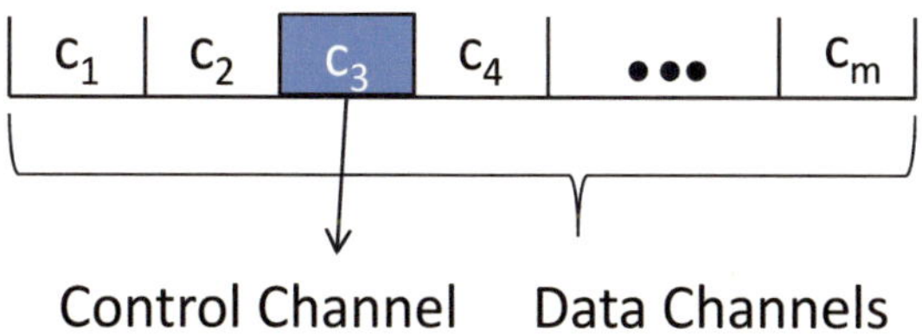

Fig. 17. a) In-band Control Channel

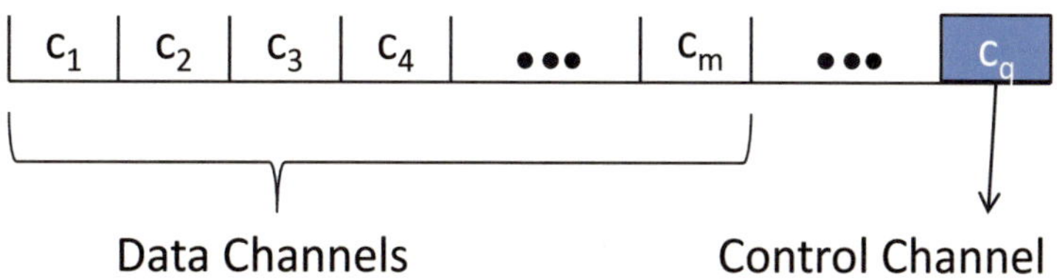

Fig. 17. b) Out-band Control Channel

Most control signals should be sent via broadcast to the users in the CRN. Several broadcasting problems such as the minimum broadcasting energy problem (Cagalj et al, 2002) and the allocation for broadcasting heterogeneous data in multiple channels (Hsu et al, 2005; Tsai et al, 2009), among others have been studied in the literature. The channel allocation/frequency assignment problem has been studied in static and dynamic environments. An overview of models and solutions of the frequency assignment problem in those environments can be found in respectively in (Aardal et al, 2007) and (Katzela & Naghshineh, 1996).

The broadcast frequency assignment problem for frequency agile networks, i.e. networks in which users can shift their operating frequency, was introduced by Steenstrup (Steenstrup, 2005). The problem is analyzed for an ad-hoc network and a Greedy approach was used to find the minimum number of channels that are needed for broadcasting information.

For CRN in general, and for heterogeneous frequency CRN, specifically, a fixed CCC might not be available. Some of the reasons could be different PU presence according to the location, for homogeneous frequency CRN, and also different sets of channels for the heterogeneous case. For solving this problem, and in order to use as minimum energy as possible, a minimum number of clusters (channels), must be found. In Fig. 18, the minimum number of channels for the example used in Fig. 7 is found.

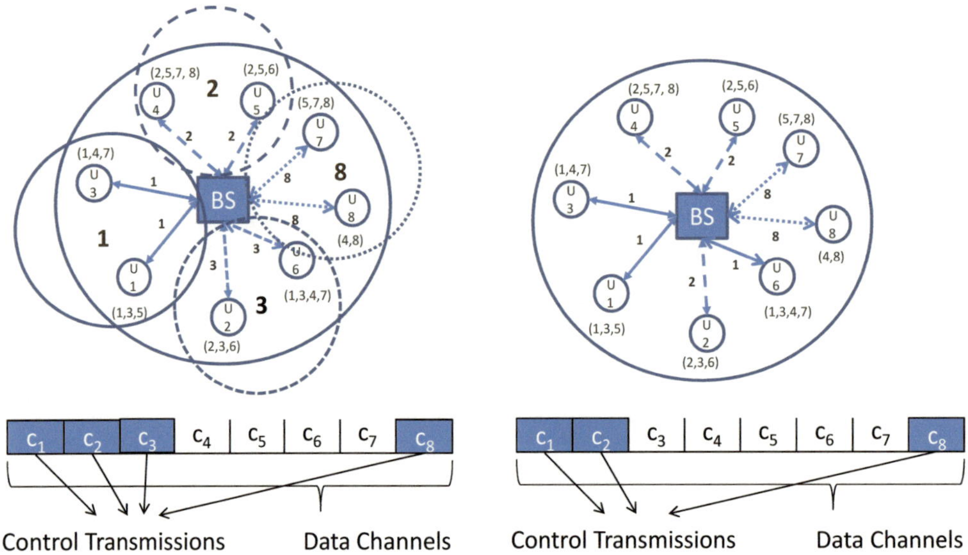

Fig. 18. Minimum number of channels for a clustered MCM

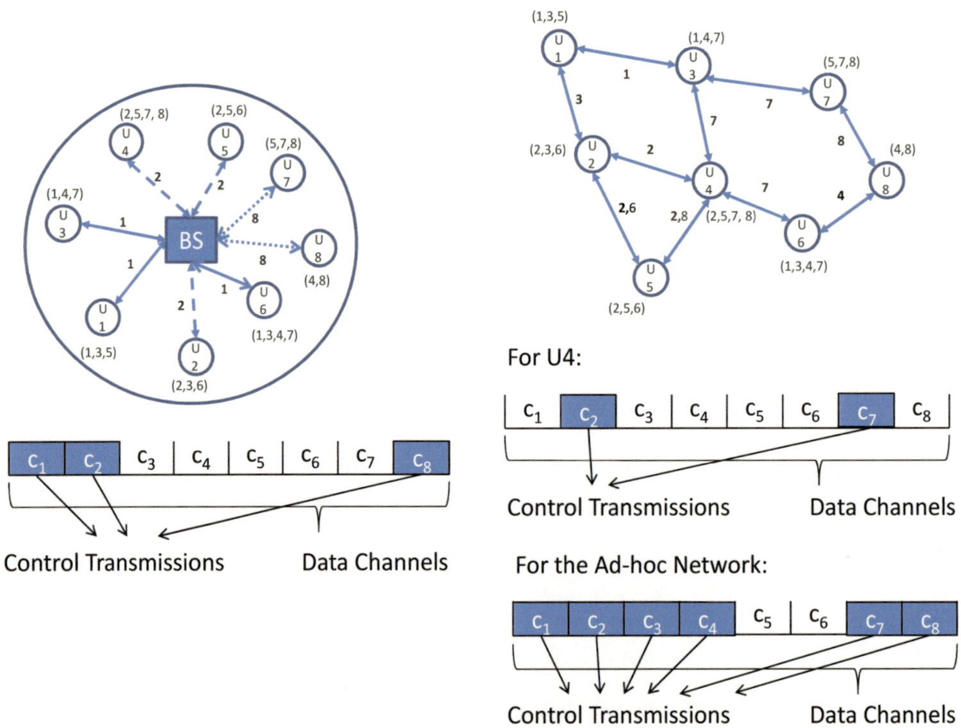

Fig. 19. Minimum Channel Problem for Centralized and Ad-hoc CRN

In (Kunar et al, 2008), the authors define the clusters for finding this minimum number of frequency channels under the same conditions used in (Steenstrup, 2005). In (Lazos et al, 2009), the authors considered the control plane and used the clustering approach for finding the minimum number of channels needed for control in a CRN. A greedy approach is used to solve the corresponding clustering problem. For future work, we plan to use several techniques for solving the *minimum number of channels problem* in both centralized and ad-hoc networks as shown in Fig. 19, using the example from Fig. 14.

In the following lines, the bases for solving for this channel allocation/frequency assignment problem are presented by implementing a combined spectrum access/mobility strategy in the control plane.

3.2 Multiple control messaging

One of the main considerations for studies in frequency assignment problems is that a channel can generate interference in adjacent channels. The authors have presented a basic model, shown in Fig. 20, for a Centralized CRN that uses CPCs for signalization and control (Bolívar et al, 2010).

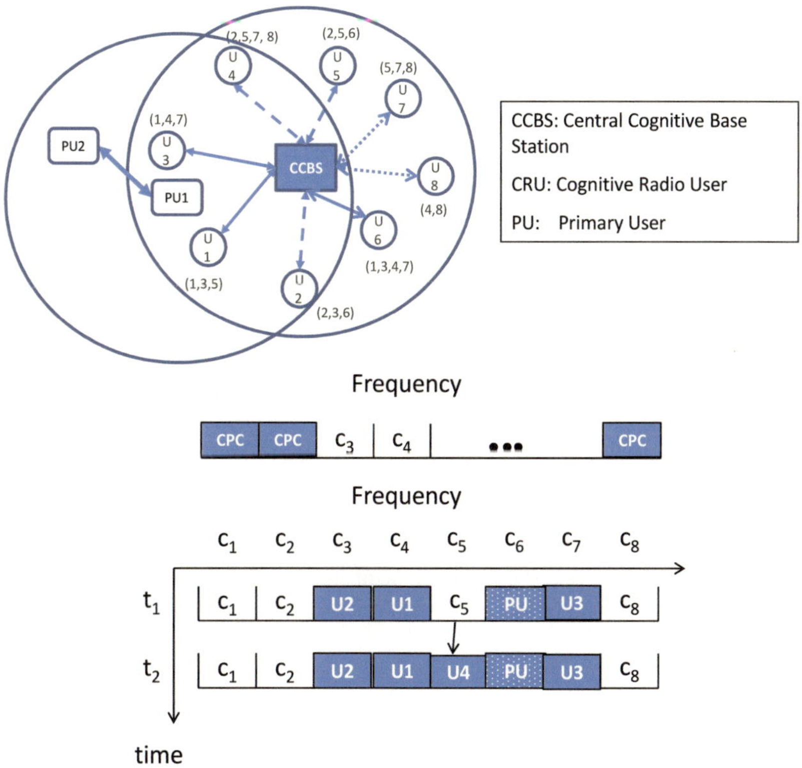

Fig. 20. Cognitive Radio Model

The main idea was to introduce a control signal, basically periodical beacons, to announce channel availability and the necessity of leaving a frequency slot if that one was occupied. In our scenario, since the broadcast signaling is transmitted the same for each channel and only in a couple of a large number of sub-channels (Bolívar et al, 2010; Bolívar & Marzo, 2010), we can assume that using adequate modulation/coding schemes, interference among adjacent channels is non-existent.

3.3 Shared control messaging

The basic model of the CRN provided control signaling through CPCs distributed in every available channel or frequency slot. The control is performed by using frequency-division and time-division multiplexing techniques, and allows the utilization of the CRN by heterogeneous CRU devices. However, in terms of energy, transmitting through every available channel would be inefficient. This is because the wireless spectrum channels would be occupied in a specific moment. Considering this problem, new alternatives should be explored to reduce the energy used for control signaling CRUs channel availability. In order to reduce the energy consumption, the authors used the characteristics of the time/frequency combined approach for the Central Cognitive Base Station (CCBS) to only signal a new available channel when a CRU that was not transmitting is requesting communication (Bolívar & Marzo, 2010). We also considered the benefits of using a distributed control and a centralized database for reducing the amount of energy used to signal this availability in the CRN. Using the example from Fig. 4, Fig.6 and Fig. 18, the SCM and MCM of this model is shown in Fig. 21.

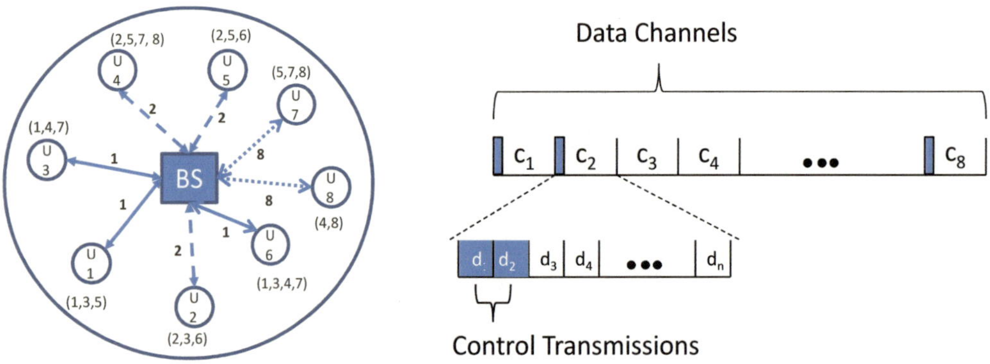

Fig. 21. Shared and multiple control messaging (Frequency-division)

3.4 Hoping control messaging

In Fig, 22, an example of the time/frequency approach is shown. According to the example in Fig. 20 and Fig. 21, U4 has four channels for communications (c_2, c_5, c_7 and c_8) and "senses" its environment.

Channel c_7 is already used by U3, so this channel is unavailable. Among the other channels, U4 decides to use c_5. Channel c_3 is occupied by U2, c_4 is occupied by U1 and c_6, by a PU. Suppose that a PU wants to use c4 in a moment t, $t_3 < t < t_4$. Using the time slot division, U1 is able to know that the channel must be evacuated and U1 starts transmitting in the following time slot in c_1.

The CCBS, however, still needs to broadcast signals to its users, especially when unexpected PU communication appears in the CRN in some specific moments. This, as expected, is a part of the spectrum mobility issue. Using the same example from Fig. 21, let's suppose that a PU that uses c_8 appears in t_i, with $t_3 < t_i < t_4$, and a PU that uses c_2 appears in t_j, with $t_5 < t_j < t_6$. We can see an approximate situation in Fig. 23.

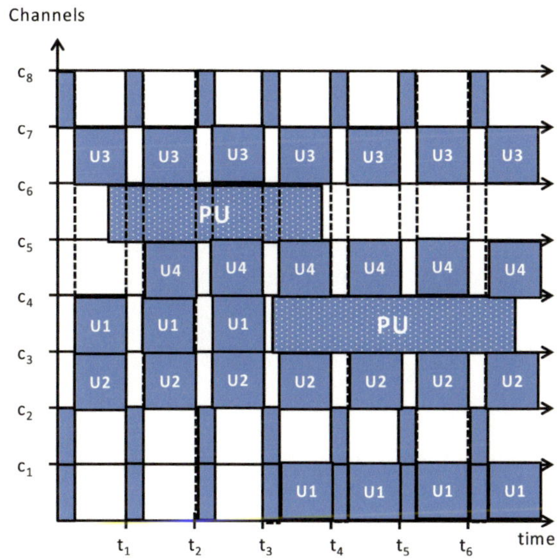

Fig. 22. Time slot utilization by both Primary Users (PU) and Cognitive Radio Users (U1, U2, U3, U4) in time

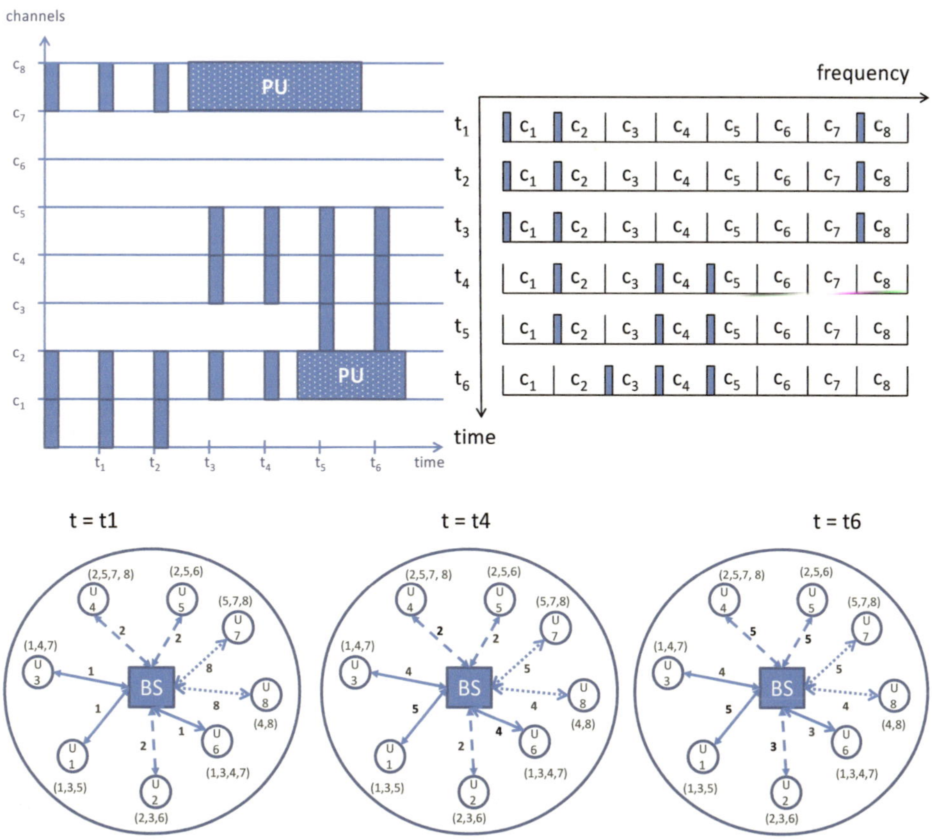

Fig. 23. Spectrum Mobility and HCM.

The control messaging must hop in t = t_4 from c_8 to another channel. However, in this process, in order to maintain the same number of channels, the control messaging from c_1 also hops. All users are covered by c_2, c_4 and c_5. In t = t_6, c_2 is unavailable, so its control transmissions are split into c_3 and c_5.

3.5 Overlay control messaging

As mentioned before, the main idea in this work is to use OSA to guarantee that no PU is interfered by a PU transmission by transmitting above a power threshold. Furthermore, we want to guarantee that when a PU is communicating, no other signal is in its same channel for security reasons. This approach is clearly seen in Fig. 23.

4. Conclusion

The control plane for Cognitive Radio Users is a very important part for the spectrum access and mobility in a CRN. However, current studies for the control transmissions are not strongly correlated. Different authors propose their methods for controlling the CRN; however, since there was not a clear classification of the control strategies, to decide which strategy is most suited to a specific CRN could be a very difficult to perform.

This is the reason why in this chapter we wanted to propose a classification for the transmission of control messages as a blueprint in order to compare the advantages and disadvantages of these control strategies. Each control mechanism can be classified according to four basic characteristics: control messaging channel dedication, number of channels used for control messaging, changes on the location of these channels over time and level of power for transmitting the control messages.

Furthermore, we study a previous model introduced in (Bolívar & Marzo, 2010) by using this classification: the control plane for a centralized CRN with heterogeneous frequency devices (HFD). In order to fulfill the basic control characteristics for spectrum access and mobility, the control strategy is presented as a combination of shared, multiple (clustered), hoping and overlay control messaging (SMHOCM).

Several concepts as the beacon strategy and CPCs are also introduced and a combined time/frequency approach is presented. We consider that the best way to control the centralized CRN with HFD is by using this SMHOCM approach. However, we encourage researchers to suggest others, by using the classification previously provided.

For future works, we would like to compare the existent control strategies in environments where all of them are suitable. Moreover, we would expand the study of the control plane for CRAHNs.

5. Acknowledgement

Part of this work was supported by the Department of Universities, Research and Information Society (DURSI) of the Government of Catalonia, European Social Funds (SGR-1202), and by a FI Grant from the Government of Catalonia, in accordance with the Resolution IUE/2681/2008, and also by the Spanish Government (TRION MICINN TEC2009 - 10724).

6. References

Aardal, K. et al. (2007). Models and solution techniques for frequency assignment problems. *Annals of Operations Research*. Vol. 153, No. 1, (May 2007), pp. 79 - 129. ISSN 0254-5330.

Akyildiz, I. F. et al. (2008). A survey on spectrum management in cognitive radio networks. *IEEE Communications Magazine*. Vol. 46, No. 4, (April 2008), pp. 40-48. ISSN 0163-6804.

Bolívar, N.; Marzo, J. L. & Rodríguez-Colina, E. (2010). Distributed Control using Cognitive Pilot Channels in a Centralized Cognitive Radio Network. *Proceedings of the the Sixth Advanced International Conference in Telecommunications*. pp. 30-34, ISBN: 978-0-7695-4021-4. Barcelona, Spain. May 9-15, 2010.

Bolívar, N. & Marzo, J. L. (2010) Energy Reduction for Centralized Cognitive Radio Networks with Distributed Cognitive Pilot Channels. *Proceedings of the IEEE Latin-American Conference on Communications 2010, Latincom 2010*. pp. 1-5. ISBN 978-1-4244-7171-3. Bogotá, Colombia. September 15-17, 2010.

Bolívar, N. & Marzo, J. L. (2011). Broadcast Signaling for a Centralized Cognitive Radio Network with Distributed Control. *Proceedings of the First International Conference on Advances in Cognitive Radio (COCORA 2011)*. pp. 42-47. ISBN 978-1-61208-131-1. Budapest, Hungary. April 17-22, 2011.

Bourse, D. et al. (2007). The E2R II Flexible Spectrum Management (FSM) Framework and Cognitive Pilot Channel (CPC) Concept - Technical and Business Analysis and Recommendations. In: *E3. White Papers E2R II*. November 2, 2011. Available from: https://www.ict-e3.eu/project/white_papers/e2r/7.E2RII_FSM_CPC_UBM_White_Paper_Final%5B1%5D.pdf.

Cagalj, M.; Hubaux, J.-P. & Enz, C. (2002). Minimum Energy Broadcast in All Wireless Networks: NP-Completeness and Distribution Issues. *Proceedings of the 8th annual international conference on Mobile computing and networking, MobiCom 02*. pp. 172-182. ISBN: 1-58113-486-X. Atlanta, Georgia, USA. September 23-28, 2002.

Chowdury, K. R. & Akyildiz, I. F. (2011). OFDM-Based Common Control Channel Design for Cognitive Radio Ad Hoc Networks. *IEEE Transactions on Mobile Computing*. Vol 10, No.2, (February 2011), pp. 228-238. ISSN 1536-1233.

Filo, M. et al. Cognitive Pilot Channel: Enabler for Radio Systems Coexistence. *Proceedings of the Second International Workshop on Cognitive Radio and Advanced Spectrum Management 2009, CogART 2009*. pp. 17-23, May 2009.

Ghasemi, A. & Sousa E. S. (2008) Interference Aggregation in Spectrum-Sensing Cognitive Wireless Networks. *IEEE Journal of Selected Topics in Signal Processing*. Vol 2, No. 1, (February 2008), pp. 41-56. ISSN 1932-4553.

Hsu, C. H.; Lee, G. & Chen, A. L. P. An Efficient Algorithm for Near Optimal Data Allocation on Multiple Broadcast Channels. *Distributed and Parallel Databases*. Vol. 18, No. 3, (November 2005), pp. 207-222. ISSN 0926-8782.

Hulbert, A. P. Spectrum Sharing Through Beacons. (2005). *Proceedings of the 16th IEEE International Symposium on Personal, Indoor and Mobile Radio Communications*. pp. 989-993, ISBN 978-3-8007-2909-8. Berlin, Germany. September 11-14, 2005.

Jing, X. & Raychaudhuri, D. Global Control Plane Architecture for Cognitive Radio Networks. *Proceedings of the IEEE International Conference on Communications 2007, ICC 2007*. pp. 6466-6470. ISBN 1-4244-0353-7. Glasgow, Scotland. June 24-28, 2007.

Katzela, I. & Naghshineh, M. (1996). Channel Assignment Schemes for Cellular Mobile Telecommunication Systems: A Comprehensive Survey. *IEEE Personal Communiactions Magazine.* Vol. 3, No. 3, (June 1996), pp. 10 - 31. ISSN 1553-877X.

Kunar, V. S.; Pemmaraju, S. V. & Pirwani, I. A. (2008). On the complexity of Minimum Partition of Frequency-Agile Radio Networks. *Proceedings of the 3rd IEEE Symposium on New Frontiers in Dynamic Spectrum Access Networks, DySPAN 2008.* pp. 1-10. ISBN 978-1-4244-2016-2. Chicago, Illinois, USA. October 14-17, 2008.

Lazos, L.; Liu, S. & Krunz, M. (2009). Spectrum Opportunity-Based Control Channel Assignment in Cognitive Radio Networks. *Proceedings of the 6th Annual IEEE Communications Society Conference on Sensor, Mesh and Ad Hoc Communications and Networks, SECON '09.* pp. 1-9. ISBN 978-1-4244-2907-3. Rome, Italy. June 22-29, 2009.

Mangold, S.; Jarosch, A. & Monney, C. (2006). Operator Assisted Cognitive Radio and Dynamic Spectrum Assignment with Dual Beacons - Detailed Evaluation. *Proceedings of the First International Conference on Communication System Software and Middleware 2006.* pp. 1-6. ISBN 0-7803-9575-1. Dehli, India. January 8-12, 2006.

Mitola III, J. & Maguire, G. Q Jr. (1999). Cognitive Radio: Making Software Radios More Personal. *IEEE Personal Communications (Wireless Communications)*, Vol.6, No. 4, (August 1999), pp. 13-18. ISSN 1070-9916.

Pérez-Romero, J. et al. A Novel On-Demand Cognitive Pilot Channel enabling Dynamic Spectrum Allocation. *Proceedings of the 2nd IEEE Symposium on New Frontiers in Dynamic Spectrum Access Networks, DySPAN 2007.* pp. 46-54. ISBN 1-4244-0663-3. Dublin, Ireland. April 17-20, 2007.

Sallent, O. et al. (2009). Cognitive Pilot Channel Enabling Spectrum Awareness. *Proceedings of the IEEE Conference on Communications Workshops 2009.* pp. 1-6. ISBN 978-1-4244-3437-4. Dresden, Germany. June 14-18, 2009.

Shukla, A. et al. (2007). Cognitive Radio Technology - A study for OFCOM. QinetiQ Ltd. January 30, 2012. Available from: http://stakeholders.ofcom.org.uk/binaries/research/technology-research/cograd_main.pdf

Steenstrup, M. E. (2005). Opportunistic Use of Radio-Frequency Spectrum: A Network Perspective. *Proceedings of the 1st IEEE Symposium on New Frontiers in Dynamic Spectrum Access Networks, DySPAN 2005.* pp 638 - 641. ISBN 1-4244-0013-9. December 2005. Baltimore, Maryland, USA. November 8-11, 2005.

Theis, N. C.; Thomas, R. W & DaSilva, L. A. (2011) Rendezvous for Cognitive Radios. *IEEE Transactions on Mobile Computing.* Vol 10, No. 2, (February 2011), pp. 216-227. ISSN 1536-1233.

Tsai, H.-P.; Hung, H.-P. & Cheng, M.-S. (2009). On Channel Allocation for Heterogeneous Data Broadcasting. *IEEE Transactions on Mobile Computing.* Vol. 8, No. 5, (May 2009), pp. 694-708. ISSN 1536-1233.

Wang, F.; Krunz, M. & Cui, S. (2008). Spectrum Sharing in Cognitive Radio Networks. *Proceedings of the 27th IEEE Conference on Computer Communications, INFOCOM 2008*, pp. 36-40. ISBN 978-1-4244-2025-4. Phoenix, Arizona, USA. April 13-18, 2008.

3

Cognitive Media Access Control

Po-Yao Huang
National Taiwan University
Taiwan

1. Introduction

With the increasing demand of wireless communication and the fact of under utilization of wireless spectrum, a lot of researches have developed technologies to enhance the efficiency of spectrum utilization. The spectrum allocation chart by the Federal Communications Commission (FCC) (NTIA, 2003) indicates high degree legislation of frequency band, in other words, our sky is quite crowded. However, another study by FCC (Report & Order, 2002) also points out the fact that the conventional, fixed spectrum allocation policy is inadequate in addressing nowadays rapidly growing wireless communication. Vast temporal and geographic variations in the usage of allocated spectrum utilization ranging are from 5% to 75% below 3GHz while the spectrum utilization is even worse in the range above 3GHz. Generally speaking, many licensed spectrum blocks are idle at most of the time. These unused spectrum blocks are known as *spectrum holes*, which is defined in (Haykin, 2005) : *A spectrum hole is a channel (i.e. frequency band) assigned to a primary user, but, at a particular time and specific geographic location, the channel is not being utilized by that primary user.* To resolve the spectrum inefficiency in the more demanding reality, dynamic spectrum access (DSA) techniques are proposed to resolve the challenge of insufficient spectrum utilization as investigated in (Akyildiz et al., 2006) and (Jha et al., 2011).

1.1 Cognitive radio networks

The key to enable DSA is cognitive radio (CR) (Mitola & Maguire, 1999) technology, which is conventionally defined as a physical/link-level technology aims at realizing DSA. In the standardization as IEEE 802.22 (IEEE, 2009), CR has been considered as a promising technology to enhance spectrum/channel efficiency while primary systems (PSs) are with relatively low spectrum utilization. CR provides the capability to share the channel with PSs' users in a opportunistic way. Moreover, CR can further provide networking "macro-scale diversity" above link layer to bridge the integrated re-configurable system networking vision as described in (Chen et al., 2008). Such a scenario for future wireless networks is recognized as cognitive radio networks (CRNs) which are envisioned to provide high bandwidth and good quality of service (QoS) to mobile users via heterogeneous wireless network architectures by reconfigurable CR transceiver for DSA. CRN can be deployed in network centric, fully distributed, Ad-hoc, and mesh architectures and serve both licensed and unlicensed applications. The basic component of CRN are model station (MS), base station/access point (BS/AP), and backbone/core networks. These components construct three kinds of network architectures in CRN : Infrastructure, ad-hoc, and mesh networks.

There are generally two kinds of wireless communication systems in CRN : primary system (PS) and cognitive radio (CR) system. They are classified by their properties on a specific channel. In general, a PS is licensed. PS is an existing system operating on the channel and has a higher priority to utilize its channel than other CRs. On the contrary, CRs, which are capable to switch over multiple channels, facing the new challenges to exploit the spectrum holes over multichannel as well as avoid interfering PSs simultaneously. These new challenges define a new media access control (MAC) problem under CR paradigm. With different PSs, the inherent characteristic in the MAC of CRN is a multichannel scenario with high priority PSs' users and competing CRs. In next section, we will first discuss the conventional multichannel MAC problem followed by the new challenge of MAC under CR paradigm.

1.2 Conventional media access control

Before introducing the MAC for CR, we start from the conventional MAC problem (Bertsekas & Gallager, 1992). As it contents, stations share a common media(i.e.,channel)and try to communicate on the shared channel. (Note that the term "station" is used in this subsection of discussion in comparison of "CR" in following sections). Without losing generality, the original MAC problem with *N* devices/stations in local area network (LAN) can be shown as Fig. 1-(a).

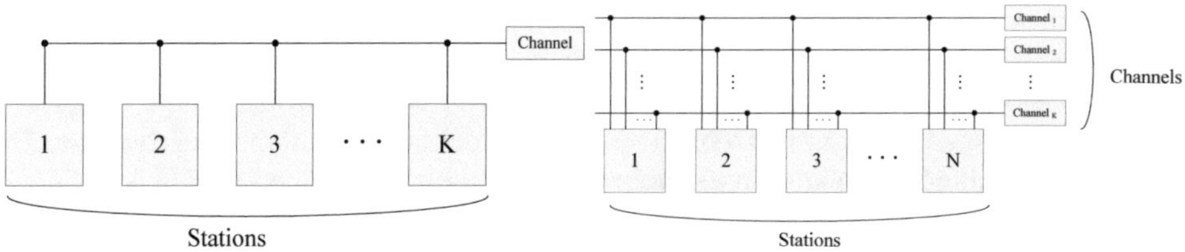

(a) Conventional MAC (b) Multichannel MAC

Fig. 1. Subfigure with four images

All stations share a common media and try to utilize the media for transmission. If more than two stations try to access the shared media simultaneously, then collision happens. In other words, there is contention between stations for the limited resource. Many protocols have been developed to resolve this problem such as CSMA/CD in local area network (LAN) and CSMA/CA in wireless LAN (WLAN). As discussed in (Marsan & Roffinella, 1983), multiple orthogonal channels provide a new degree of freedom to further alleviate the contention. Multichannel MAC, depicted as Fig. 1-(b), is considered to utilize multiple orthogonal channels for throughput enhancement by allowing parallel transmissions over channels. The conventional design of multichannel MAC under wired LAN is simple, stations equipped with multiple volt-meters may tuned to access those multichannel(separated wires or orthogonal frequency bands). Collision still happens if there is a co-channel transmission simultaneously.

A more critical and challenging scenario lies in the multiple wireless networks, such as dynamic spectrum access over multiple wireless communication systems shown in Fig. 2. The main difference between wired and wireless networks is that now stations cannot access multiple systems simultaneously under hardware design restrictions. At a specific time period, station can only select and switch to one specific channel for communication. Therefore, the wireless multichannel MAC problem is further been divided into two parts.

The first part is "Collision Avoidance/Resolution." Collision avoidance/resolution (Sun et al., 1997) is inherited from conventional single channel MAC problem. Contention must be avoided/resolved when more than two stations try to access the channel. Stations must avoid simultaneous transmission and resolve further possible failure after collision happens. On the other hand, with the limited hardware capability for channel access, a stations may listen/transmit over one selected channel. Therefore, a scheme/protocol framework must be properly designed for multiple stations to select the proper channel for data transmission. To achieve so, the brand new problem lies in wireless multichannel MAC is the "Channel Selection" part. Channel selection, which considers distributed selection of communication channel, is the new challenging issue which recently attracts most efforts as in (Mo et al., 2008).

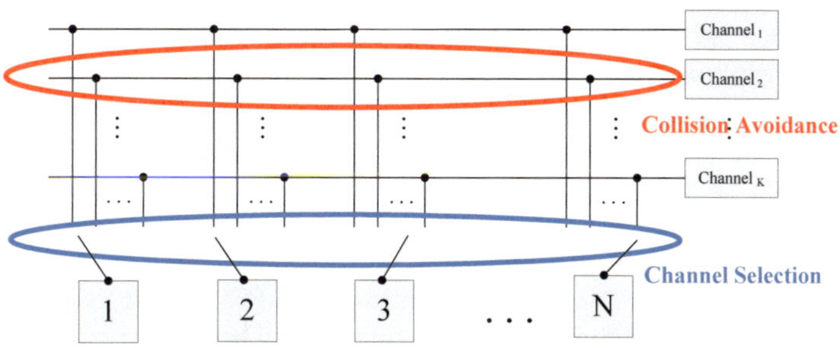

Fig. 2. Two fundamentals of Multichannel MAC : 1) Collision Avoidance/Resolution and 2)Channel Selection.

1.3 Cognitive media access control for CR

The generalized multichannel MAC under CR paradigm can be illustrated in Fig. 3. In CRN, CRs are categorized as secondary users who sense for the spectrum/channel opportunity and access the channel with a lower priority than PSs' users. Since CR and PS belong to different systems, the feasible multichannel MAC protocol for CR is required to resolve not only intra-system (CR-CR,within CRN) but also inter-system (CR-PSs) media contention. Under the design limitation that each CR cannot access all of the available channels simultaneously (i.e., CR equipped with only one reconfigurable transceiver/radio), the main task of CR MAC is to distributively choose the channel to use without a reserved common control channel.

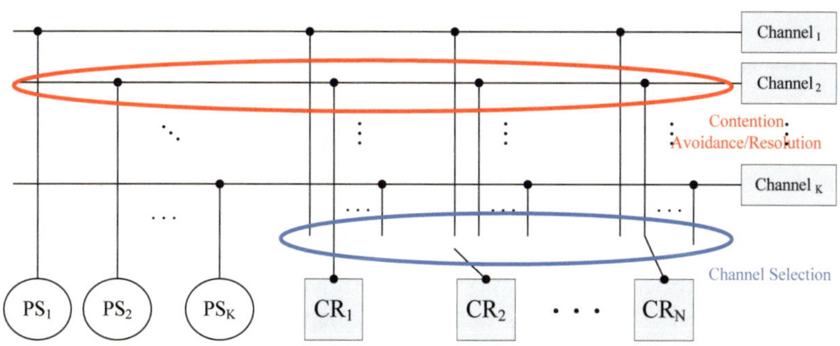

Fig. 3. CR multichannel MAC

Based on the discussion above, CR multichannel MAC differs from conventional wireless multichannel MAC problem. CRs, as secondary users, must agilely avoid interference with prior communication users. Moreover, for support of multiple access among CRs, a contention avoidance/resolution mechanism must be properly designed. To further enhance the spectrum efficiency and throughput of CRN, a feasible channel access strategy should also be properly designed. Therefore, the aims of CR multichannel MAC are :

1. Efficiently utilizing multichannel by initiating parallel transmissions in a distributive manner.
2. Meeting the requirement that transmissions of CRs are interference-free for PSs.
3. Alleviating contention over CRs and PSs, maximizing the CRN throughput.

Moreover, CR, which is envisioned to be equipped with sensing and cognitive capability, brings new possibility for further performance enhancement under its challenging multiple co-existing PSs MAC scenario. CR, with ability to retrieve environmental information and learn to adapt itself, is possible to achieve the optimal performance. For example, the stabilization under CR multichannel MAC is a more complicated problem then conventional single channel solution (Limb, 1987). Take stabilization as an example, 1) Intra-system (CR-CR, within CRN) information : the load distribution over multiple channels and 2) Inter-system (CR-PSs) information behaviour of PSs can be retrieved by cognitive functionality of CR for further throughput optimization. The related cognitive functionality will be introduced, analysed and discussed in the following sections.

In this chapter, we will develop a general protocol scheme for CR multichannel MAC. A CSMA-based cognitive MAC is proposed with a general Markov Chain analysis and the steady state approximation for practical design insight. The cognitive functionality of CR and optimization is also proposed for MAC protocol optimization. The simulation results validate the analysis, and are compared with the results with current research works.

This chapter is organized as follows Section II summaries the related work. Section III describes the system model under consideration. The proposed general CR multichannel MAC scheme is in Section IV with a CSMA-based cognitive MAC design. In Section V, we model, analyse and provide a steady-state protocol performance approximation. Cognitive functionality and performance enhancement and optimization is introduced in Section VI. Section VII is the simulation results and discussion. Conclusion and future works are presented in section VIII.

2. Multichannel MAC: theory and related works

Representative multichannel MAC protocols are summarized in Table 1. These protocols can further be categorized as narrow-band approaches and wide-band approaches. For narrow-band transmitter and receiver, the frequency band to be used for transmission is to be predetermined, or dynamically chosen. While in wide-band system, the transmitter can simultaneously transmit over multiple frequency bands that are detected to be unoccupied while the receiver can retrieve information over multichannel simultaneously. Note that in wide-band systems, each station may need more than two radios to achieve the parallel transmissions over multichannel while in narrow-band system, a station is equipped with only one radio. (Two radios in dedicated control channel multichannel MAC case.) These multichannel protocols can further be categorized as follows:

The Related Works		
Category	Protocol Type	Reference (Year
Narrow-Band	Dedicated Control Channel	(Wu et al., 2000) (Zhang & Su, 2011) (WC Hung, 2002) (Cordeiro & Challapali, 2007) (Liu & Ding, 2007) (Timmers et al., 2007)
	Common Hopping	(Tang & Garcia-Luna-Aceves, 1998) (Tzamaloukas & Garcia-Luna-Aceves, 2000)
	Split Phase	(Chen et al., 2003) (So & Vaidya, 2004)
	Multiple Rendezvous	(Bahl et al., 2004) (Mo et al., 2008)
Wide-Band	CDMA/Frequency Coding	(Al-Meshhadany & Ajib, 2007) (Zhang et al., 2003)
	Multi-Radios	(Nasipuri et al., 1999) (Jain et al., 2001)

Table 1. Related Works

2.1 Protocols with dedicated control channels

As shown in Fig. 4-(a), one (or more) control channels are utilized to exchange control signal (such as request to send (RTS) and clear to send (CTS) in WLAN) and distributed sensing (Liu & Ding, 2007) or cooperative sensing (Zhang & Su, 2011) information of CRs. CR using this type of protocol must equip with at least two radios, one (or more) radio is reserved to listen to the control channel. It should be noted that the reservation of a common control channel violates CR design philosophy which utilize the spectrum holes in a opportunistic sense. The performance bottleneck of this kind of protocol is the overhead over the reserved common control channel.

2.2 Protocols with common hopping

In protocols with common hopping, CRs hop across multichannel following the same hopping pattern (Tang & Garcia-Luna-Aceves, 1998),(Tzamaloukas & Garcia-Luna-Aceves, 2000). As shown in Fig. 4-(b), common hopping is comparable to "switch" control channels over multichannel. The bottleneck of such protocol type is still the overhead of the hopping control channel. This type of protocol requires only one radio.

2.3 Split phase protocols

Depicted in Fig. 4-(c), time is divided into control and data phases (Chen et al., 2003), (So & Vaidya, 2004). CR exchange control information on a dedicated control channel during the control phase, then accessing the channel during the data phase. The idea is analogous to "reserve" some time duration as control channel. The bottleneck is still the overhead of the control channel. One radio is required using this type of protocol.

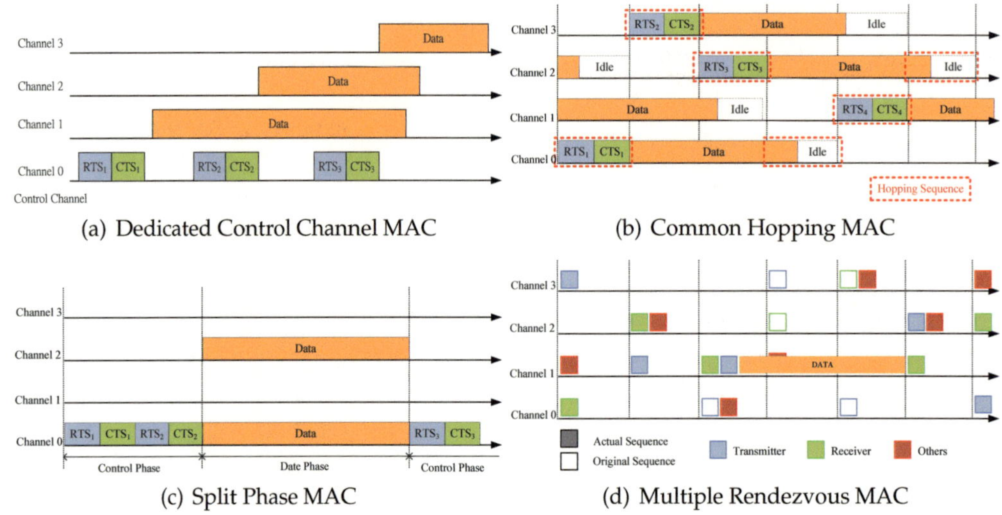

Fig. 4. Related Multichannel MAC Protocols

2.4 Multiple rendezvous protocols

In (Bahl et al., 2004) and (Mo et al., 2008), CRs opportunistically select a channel, exchange control information then transmit data. Shown in Fig. 4-(d), the main idea is comparable to "separate" control channel over multichannel. Multiple transmissions can be ignited over channels with only one radio. Parallelism is the design philosophy of this kind of protocol. However, the cons of these results are the lack of joint consideration of channel selection and collision avoidance/resolution of inter-system (PS-CR) and intra-system (CR-CR). The performance is degraded under PSs presentation and contention over CRs.

From the discussion above, one should note that for the feasible CR realization, the preserved common control channel should be avoided since it violates the design philosophy of CR. Furthermore, CR with only one transceiver is more practical design. In the following section, we provide a general model for analysis of multichannel MACs and propose the cognitive multichannel MAC which optimizes CRN throughput performance by enabling multiple rendezvous and fits requirements of multichannel MAC under CR paradigm.

3. System model

3.1 Network model

The considered generalized CR multichannel MAC is a fully-distributed ad-hoc network as in (Mo et al., 2008). The infrastructure scenario under CRN paradigm is exampled and illustrated in Fig. 5. There are multiple co-existing PSs, while CRs tries to utilize multichannel to communicate with other CRs or base station of CRN. Physically, CRs are capable to reconfigure its physical layer to access every channel selected from the pool of multichannel where different PSs pre-occupy. Categorized as secondary users, CRs seek channel opportunities over multichannel and access the channel to relay its data to the backbone network or other CRs without interfering PSs' users.

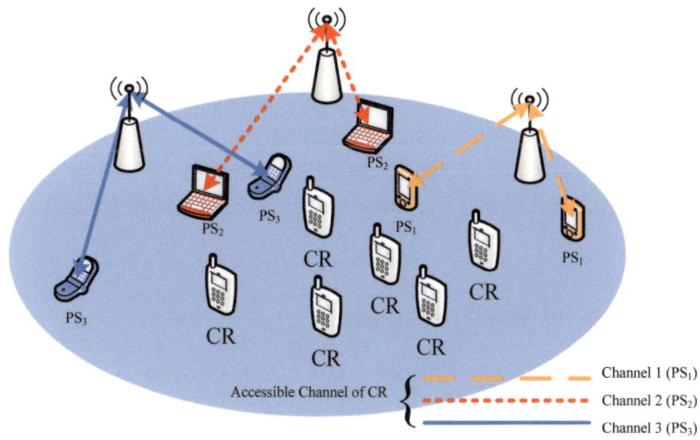

Fig. 5. CR System Model

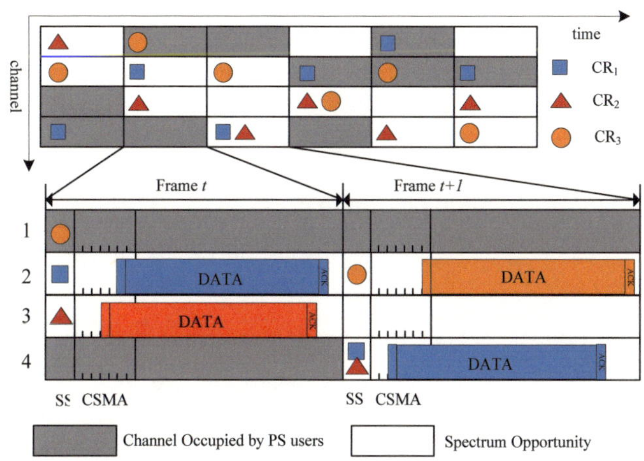

Fig. 6. Cognitive CSMA-based McMAC protocol for CRN

3.2 Multichannel MAC under CR paradigm

The general CR multichannel MAC is depicted in Fig. 6. Time is divided into small synchronized (t as the frame index) slots over multichannel. The duration of a slot/frame is denoted as t_l. The CR multichannel MAC is characterized by: $\{K, N, C, \mathbf{q}\}$. K is defined as the number of multiple channels, we use the set $\{k\}$ as the pool for channel selection. The switching of channel take place only at slot boundary. Each channel k is supposed to be independently pre-occupied by some licensed PS or by CR. CR access the channel only if there is no PS appearance and vice versa. The appearance of PS on channel can be represented as a on-off two-state Bernoulli process with parameter q_k. We define $\mathbf{q} \triangleq \{q_k | k = 1, 2 \ldots K\}$ as the set of probabilities of PSs' appearance over channel $1 \ldots k$.

The number of CRs is denoted as N. The value of N can be estimated by neighbour discovery, network initialization algorithms as in traditional study of Ad-Hoc network as in (RoyChoudhury et al., 2000) and (Jakllari et al., 2005). Through this chapter we assume this value is correctly estimated by CRs in a distributed manner with no transient behaviour. The average capacity over channels which indicates the maximum throughput per slot is denoted as C.

The performance indicator for CR multichannel MAC paradigm are 1) the aggregated throughput R, defined as the aggregated physical layer throughput over all channels, and the 2) average channel utilization U, defined as the average number of channels being successfully utilized for data transmission, normalized by the number of channel. Note that U is equivalent to traditional MAC "throughput" defined as the number of successful transmission per channel per slot. The notations of generalized CR multichannel MAC are summarized in Table 2.

4. General CR multichannel MAC protocol

There are two components in multichannel MAC: 1) Channel Selection and 2) Contention Avoidance/Resolution. The general CR multichannel MAC and its frame structure is illustrated in Fig. 6. At the beginning of a frame, CR selects a channel with the channel selection strategy, denoted as Γ . In the frame operation, the contention avoidance/resolution is composed of two parts: 1) spectrum sensing for inter system (CR-PS) contention avoidance/resolution; and 2) intra system (CR-CR) contention avoidance/resolution mechanism. The mathematical form of these process can be represented by a contention avoidance/resolution function $g_k(n)$. With feasible channel selection and contention avoidance/resolution mechanisms, CR proceed reliable transmission.

4.1 Contention avoidance/resolution algorithm $g_k(n)$

The general form of contention avoidance/resolution can be represented by a resolution function $g_k(n)$ defined as:

$$g_k(n) = Prob(S_k = 1|n \text{ CRs attempt to transmit over channel } k) \tag{1}$$

S_k is the indicator that channel k is successfully utilized by CR. The resolution function $g_k(n)$ can further be divided into two parts: 1) the availability ϕ_k of channel k, that is, the channel must be free form PSs' users. 2) The contention resolution mechanism among CRs. Define $1_k(t)$ as the indicator denoting the availability of kth channel at slot t:

$$1_k(t) = \begin{cases} 1, & \text{channel } k \text{ is available at slot} t, \\ 0, & \text{otherwise.} \end{cases} \tag{2}$$

Spectrum sensing is performed to acquire $(1_k(t))$ of CR's currently locating channel . As a detection problem, let P_f and P_d are probability of false alarm and probability of detection, respectively. Therefore, $\phi_k = (1 - q_k)(1 - P_f^k) + q_k(1 - P_d^k)$, where ϕ_k is defined as the channel availability. For simplicity of MAC analysis, we assume perfect spectrum sensing, that is, $P_f = 0$ and $P_d = 1$, Therefore, $\phi_k = 1 - q_k$. Link the indicator function and the channel availability ϕ_k, we have:

$$\begin{cases} Prob(1_k = 1) = \phi_k = 1 - q_k \\ Prob(1_k = 0) = 1 - \phi_k = q_k \end{cases} \tag{3}$$

After spectrum sensing, contentions among CR are further alleviated with contention avoidance/resolution mechanisms. Many conventional works have been proposed to resolve this problem such as 1-persistent CSMA, p-persistent CSMA, non-persistent CSMA, etc. The contention avoidance/resolution in this part can be slotted or non-slotted. In the later section, a non-persistent CSMA mechanism is proposed and analysed.

Note	Description	Note	Description
K	Number of Channel	N	Number of CR
$\mathbf{C}$	Set of Channel Capacity	$\mathbf{q}$	Set of Prob. of PS Appearance
C	Average Channel Capacity	k	index of channel
t	Frame Index	t_l	Frame Length
q_k	Prob. of PS on Channel k	p_k	Prob. of Selecting Channel k
Γ	Channel Selection Strategy	R	Aggregated CR Throughput
U	Utilization	X_t	# of CR Transmission at slot t
S	State Space of X	i	# of current CR Transmission
j	# of CR Starting/Termination	P_{ij}	Transition Prob.
$\mathbf{P}$	Transition Prob. Matrix	π_i	Limiting Prob. of X
$\mathbf{\Pi}$	Limiting Prob. Vector	S_i^j	Prob. of Starting j Transmission
S_i^j	Prob. of Terminating j Transmission	t_l	Slot Length
s_k	Duration of Spectrum Sensing	t_k	Duration of Control Message
η_k	Protocol Efficiency over Channel k	ϕ_k	Availability of Channel k

Table 2. Notation of Generalized Multichannel MAC for CRs

Category	Protocol Name	Selection Algorithm
Deterministic	Best Channel	$p_k = 1$ where $k =$ argmax ϕ_k
Random	Uniform Selection	$p_k = \frac{1}{K}$
General	Proportional Selection	$p_k = \frac{\phi_k}{\sum_i \phi_i}$
	Optimal Selection	$p_k = p*$ (Section 6)

Table 3. Channel Selection Algorithms

4.2 Channel selection algorithms Γ

Without a reserved control channel for exchanging control information, CR distributedly selects multichannel with a channel selection algorithm Γ. Its effect can be analogous to separate CR traffic over multichannel. The resulting contention over a specific channel is thus alleviated. The general channel selection algorithm is defined as $\Gamma = \{p_1, p_2, ..., p_K\} = \{p_k\}$ and $\sum_k p_k = 1$, where p_k is the probability of selecting channel k. The category and details of channel selection algorithms are summarized in Table 3.

4.3 Protocol description

As shown in Fig. 6. With swift channel opportunity, transmission are initiated and terminated within one frame in avoidance of further interference with PSs. Moreover, With the cognitive functionality powered by sensing and adaptation of CRs, the cognition of q_k and N can be further utilized in MAC protocol optimization. The proposed cognitive multichannel MAC protocol are stated as below:

1. (Channel Selection) If CR attempts to transmit, at the beginning of a frame, the channel k is selected with probability p_k.
2. (Spectrum Sensing Phase) Perform spectrum sensing to detect the appearance of specific PS on its currently locating channel (k) and update the probabilities of PS appearance $\mathbf{q}$.
   (a) If PS appears, update q_k and skip step (3) and (4).
   (b) Else, update q_k and proceeds.

3. (CSMA-contention Phase) CR transmitter pick a back-off value between [0, N_{cw}], where N_{cw} is the contention window size. CR transmitter decrements its back-off value by one during each idle slot. In this phase, CR keeps listening to the channel.
   (a) If the channel becomes busy before the back-off value reach 0, gives up transmission.
   (b) Else, sets access indicator ($S = 1$), access the channel.
4. (Data transmission Phase) Transmit data. Transmission terminated before end of frame.
   (a) If($S = 1$), Transmit data; terminates after ACK.
   (b) Else if the data is correctly received, transmits ACK.
   (c) Otherwise, remains silent until the end of frame.

5. Protocol performance evaluation

The performance of the proposed general protocol is evaluated with Markov chain and a steady-state approximation to build intuition of parameter dependence for further optimization process in practical realization . Supposed that that CR are synchronized with PSs and spectrum sensing is perfectly performed (i.e., if PS appears at frame t, all the CRs who tuned the channel detect its appearance and avoid channel access.) The packet arrival of CR is assumed to follows a Poisson Process with parameter λ. The retransmission policy follows a geometric back-off with parameter q_r. If CR is backlogged, with probability q_r it attempts channel access to retransmit the backlogged packet.

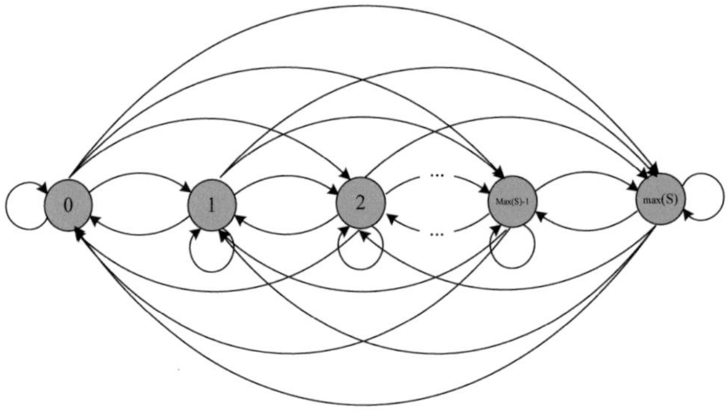

Fig. 7. The Markov Chain Model for multichannel MAC

5.1 The Markov chain model

Let X_t denote the number of backlogged CRs at the beginning of a given frame t. X_t changes from frame to frame and $\{X_t\}$ forms a discrete time Markov chain as in Fig. 7. The state space of $\{X_t\}$ is defined as: $\mathbb{S} = \{0, 1, ..., N\}$ Let $Q(x,i)$ be the probability that i un-backlogged CRs attempt to transmit packets in a given frame, and that $Q_r(y,i)$ be the probability that i backlogged node transmit. We have:

$$Q_a(x,i) = \binom{N-i}{x} (1-q_a)^{N-i-x} q_a^x \tag{4}$$

and

$$Q_r(y,i) = \binom{i}{y} (1-q_r)^{i-y} q_r^y \tag{5}$$

Where $q_a = 1 - e^{-\lambda t_l}$ is the probability of packet arrival for un-backlogged CR. The number of CR attempting channel access is therefore $x + y$ with probability $Q_a(x,i)Q_r(y,i)$.

With a smaller group on each channel for channel access, contention among CR may further be alleviated by separation of CR traffic. We define the successful transmission function $\Omega(w,a,k)$ where its value represents the probability that there are w successful transmission CRs out of a concurrent attempting CRs over channel $1,2,...,k$. With the channel selection algorithm $\Gamma = \{p_k\}$, we can find the probability that w out of a successfully wins channel contention as:

$$\Omega(w,a,k) = \sum_{n=0}^{a} \Omega(w,a-n,k-1)p_k^n(1-g_k(n)) + \Omega(w-1,a-n,k-1)p_k^n(g_k(n)) \quad (6)$$

$\Omega(w,a,k)$ can be recursively solved: Note that $\Omega(1,n,0) = g_0(n)$ and $\Omega(0,n,0) = 1 - g_0(n)$. The contention avoidance/resolution function $g_k(.)$ over channel k depends on the spectrum sensing mechanism (represented as channel availability ϕ_k) and CSMA contention mechanism. Let N_{cw} denote the contention window size, then $g_k(.)$ can be therefore derived as:

$$g_k(n) = \phi_k \sum_{i=0}^{N_{cw}-1} \binom{n}{1} (\frac{1}{N_{cw}})(1 - \frac{i+1}{N_{cw}})^{n-1} \quad (7)$$

Given x new arrival, y retransmission attempt, the probability of w successful transmission can be represented as $\Omega(w,x+y,K)Q_r(y,i)Q_a(x,i)$. The number of remaining backlogged CR in current frame is $i-y$, the number of backlogged CR after transmission attempt is $x+y-w$. For the resulting state j, the relation $j = x - w + y$ holds. Therefore, the transition probability of $\{X_t\}$ can be derived as:

$$P_{ij} = \sum_{\substack{x+w+i=j \\ 0 \le x \le N-i, 0 \le y \le i, w \le x+y}} \Omega(w,x+y,K)Q_r(y,i)Q_a(x,i) \quad (8)$$

With the transition probability matrix $\mathbf{P} = \{P_{ij}\}$, we can solve the limiting probability $\mathbf{\Pi} = \{\pi_1, \pi_2, ..., \pi_{\max(S)}\}$ of the Markov chain by $\mathbf{\Pi} = \mathbf{\Pi P}$ and $\sum_i \pi_i = 1$. The throughput can further be derived with the limiting probability $\mathbf{\Pi}$, First we define the protocol efficiency η as the portion that the MAC protocol utilizes channel capacity in a frame. Note that the frame duration is t_l; the duration for spectrum sensing is t_s; further access control (CSMA) over the channel requires a duration t_c. Therefore, $0 \le \eta \le \frac{t_l - t_s - t_c}{t_l}$. Let b denote the number of backlogged CR and π_b as its related limiting probability. The system throughput, defined as the aggregated bit transmitted over multichannel per frame can further been derived as in Eq.(9) and in Eq.(10), respectively.

5.2 Steady state approximation

Markov chain model suffers from the fact it provides no clear mathematical or physical meaning. Without losing intuition of protocol design and the hope of finding dependency among parameters for further performance enhancement, we apply a steady-state expectation approach to provide design insight. To consider the potential throughput of multichannel MAC, we analyse multichannel MAC under saturated load, controlled by Pseudo-Bayesian algorithm (Bertsekas & Gallager, 1992) with parameter p.

$$R = \sum_{b=0}^{N} \sum_{j,x,y;0\le x\le N-i,0\le y\le i,j\le x+y} C\eta j\Omega(j,x+y,K)Q_a(x,b)Q_r(y,b)\pi_b \tag{9}$$

$$U = \frac{1}{K}\sum_{b=0}^{N} \sum_{j,x,y;0\le x\le N-i,0\le y\le i,j\le x+y} j\Omega(j,x+y,K)Q_a(x,b)Q_r(y,b)\pi_b \tag{10}$$

Define X_i as the indicator of transmission attempt of CR$_i$; K, A, B as the random variables denoting the selected channel, simultaneous transmitters (including CR$_i$), and other co-channel ($K = k$) transmitters in the same operating frame, respectively. For CR$_i$ to successfully transmit its packet, the following conditions must hold: 1) CR$_i$ decides to transmit. ($X_i = 1$.) 2) The selected channel ($K = k$) must be free from PS appearance. 3) CR$_i$ wins CSMA contention ($S = 1$) with other $A = a$ out of $B = b$ co-channel transmitters. Let $P_{succ}^{(i)}$ denotes the probability of successful transmission for CR$_i$. $P_{succ}^{(i)}$ is thus conditioned on $\{S, B, A, K, X_i\}$:

$$\begin{aligned} P_{succ}^{(i)} &= \sum\nolimits_{a,b,k} P[S=1, B=b, A=a, K=k, X_i=1] \\ &= \sum\nolimits_{a,b,k} \cdot P[S=1|B=b, A=a, K=k, X_i=1] \\ &\quad \cdot P[B=b|A=a, K=k, X_i=1] \\ &\quad \cdot P[A=a|K=k, X_i=1] \\ &\quad \cdot P[K=k|X_i=1] \\ &\quad \cdot P[X_i=1] \end{aligned} \tag{11}$$

At saturated load, CR is always queued with packet. Therefore, the probability that CR$_i$ attempts to transmit is p. It is clear that,$P[X_i = 1] = p$. The probability that CR$_i$ selects channel k is $P[K = k|X_i = 1] = p_k$.

All CRs independently make access decision. The probability that there are a simultaneous transmission attempts (including CR$_i$) in the same operating frame is

$$P[A = a|K = k, X_i = 1] = \binom{N-1}{a-1} p^{a-1}(1-p)^{N-a}. \tag{12}$$

Given CR$_i$ is at channel k attempting transmission. The probability that there are exactly b co-channel CR transmission attempts out of $a - 1$ (excluded CR$_i$) simultaneous transmission attempts in the same operating frame is

$$P[B = b|A = a, K = k, X_i = 1] = \binom{a-1}{b} (\frac{1}{M})^b (1 - \frac{1}{M})^{a-b-1}. \tag{13}$$

For CR$_i$ to ''win" the channel in CSMA contention, the channel it selected must be free from PS occupation and the backoff time CR$_i$ picks must be the smallest one of other b CR. Thee contention window size is denoted by N_{cw} and we can derive:

$$P[S=1|B=b,A=a,K=k,X_i=1] = (1-q_k)\sum_{n=0}^{N_{cw}-1}(\frac{1}{N_{cw}})(1-\frac{n+1}{N_{cw}})^b. \quad (14)$$

Then the normalized throughput U / multichannel frame utilization is:

$$\begin{aligned} U &= \sum_i P_{succ}^{(i)} \\ &= N\sum_{a,b,k} P[S=1,B=b,A=a,K=k,X=1]. \end{aligned} \quad (15)$$

The second equality holds since all CRs follows a symmetric behaviour and we can remove the subscript i. It should be noted that the multichannel frame utilization is comparable to the normalized throughput in convention single channel MAC which is defined as average packet successfully transmitted frame/slot (per channel). The aggregated throughput R is:

$$R = N\sum_{a,b,k}\eta_k C_k P[S=1,B=b,A=a,K=k,X_i=1]. \quad (16)$$

6. Cognitive functionality and MAC protocol optimization

CR brings new cognitive functionality for agile DSA. In CR multichannel MAC, the sensing capability of CR for retrieving environmental information and adaptation function among CRs open new degree of freedom in MAC protocol optimization. In this section, we return to the fundamentals of multichannel MAC problem under CR paradigm: 1) Collision avoidance/resolution 2) Channel Selection and develop related cognitive functionality of CR for MAC protocol optimization.

6.1 Optimal collision avoidance/resolution for CR

CSMA-based protocol suffers from its unstable nature, a stabilization mechanism is required and will significantly contribute to performance improvement. When there are too many CRs, every CR shall decrease its potential channel access in prevention of collision with other CRs, while CR shall increase the probability of channel access when there are much resource left.

From previous sections, N and $\mathbf{q}$ can be acquired from the *sensing* function of CR in the long run; M, N_{cw}, $\mathbf{C}$ and η are predetermined during CR design. Therefore, the throughput optimization in the proposed multichannel MAC protocol can be done by finding the optimal probability of transmission attempt p^* that maximizes the aggregated throughput. Rewrite Eq.(16),

$$R = \sum_{a=1}^{N} G(a)p^a(1-p)^{(N-a)} \quad (17)$$

Where:

$$\begin{aligned} G(a) &= \sum_{b=0}^{a}\sum_{k=1}^{M}\sum_{n=1}^{N_{cw}-1}\eta_k C_k(1-q_k)(\frac{1}{N_{cw}})(\frac{N_{cw}-n+1}{N_{cw}})^b \\ &\cdot \binom{N-1}{a-1}\binom{a-1}{b}(p_k)^{b+1}(1-p_k)^{a-b-1} \end{aligned} \quad (18)$$

It should be noted that $G(a)$ is a non-decreasing function of a. In comparison to conventional single channel stabilization (Limb, 1987), information over diverse multichannel for both inter-system (CR-PSs) and intra-system (within CRN) is required to achieve stabilization over multichannel. Information retrieval can be done with the *sensing* function. The optimization problem can thus be formulated as :

Maximize

$$R = \sum_{a=1}^{N} G(a)p^a(1-p)^{(N-a)} \tag{19}$$

Subject to

$$0 \leq p \leq 1$$

Since the optimization parameter $p \in [0,1]$, Eq.(17) can be easily transformed and solved by interior-point method (Boyd & Vandenberghe, 2004) to find the optimal probability of transmission attempt p^*. Another instinct method is to construct a look-up table. This table can be pre-determined and embedded in the design of CR. By periodically adapting (*adaptation* function) the probability of transmission attempt p to $p*$, the expected throughput of CRs is then optimized. The proposed multichannel protocol is therefore optimized by *cognitive* functionalities - *sensing* function (inter-system information (q_k) and intra-system information (N) retrieval) and *adaptation* function (adapt to fit the environment.)

6.2 Optimal channel selection for CR

For N CRs distributedly follows a selection algorithm $\Gamma = \{p_k\}$ for selecting K channels for channel access, the optimization problem follows the similar method as the stabilization mechanism. Therefore, the throughput optimization in the proposed multichannel MAC protocol can be can be done by finding the optimal p_k^* that maximizes the expected throughput. Rewrite Eq.(9), we have:

$$R = \sum_{k-1}^{K} \sum_{a=1}^{N} \sum_{b=0}^{a} G'(a,b)F(k)p_k^{b+1}(1-p_k)^{a-b-1} \tag{20}$$

Where:

$$F(k) = \eta_k C_k \phi_k \tag{21}$$

$$G'(a,b) = p^a(1-p)^{(N-a)}\binom{N-1}{a-1}\binom{a-1}{b}\sum_{n=1}^{N_{cw}-1}\left(\frac{1}{N_{cw}}\right)\left(\frac{N_{cw}-n+1}{N_{cw}}\right)^b \tag{22}$$

The optimization problem can thus be formulated as follows:

Maximize

$$R = \sum_{k-1}^{K} \sum_{a=1}^{N} \sum_{b=0}^{a-1} G'(a,b)F(k)p_k^{b+1}(1-p)^{a-b-1} \tag{23}$$

Subject to

$$\sum_k p_k = 1 \text{ and } 0 \leq p_k \leq 1$$

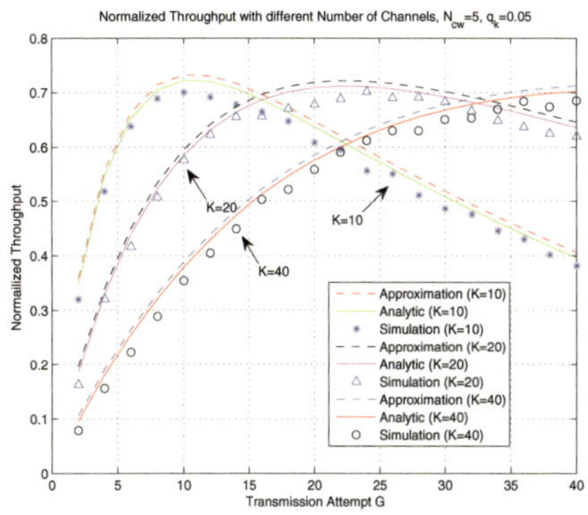

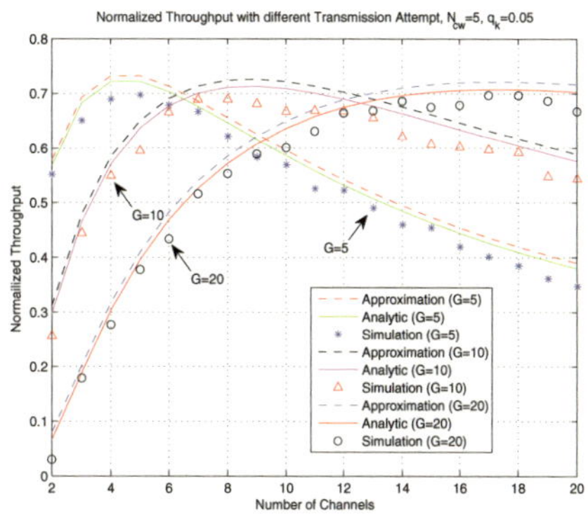

(a) Normalized Throughput U with Different Number Channel K via Transmission Attempt G, $N_{cw} = 5, q = 0.05$

(b) Normalized Throughput U with Transmission Attempt G via Different Number of Channel K, $N_{cw} = 5, q = 0.05$

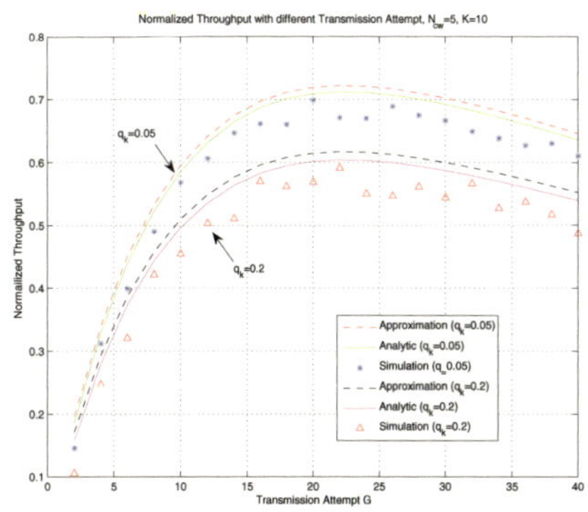

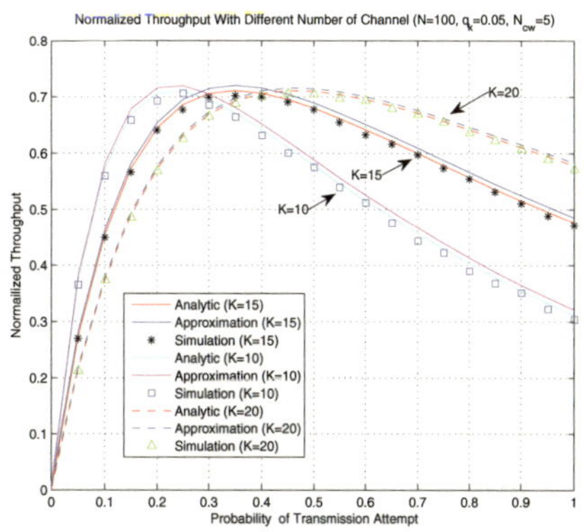

(c) Normalized Throughput U with Different Probability of PS appearance q_k, $N_{cw} = 5$, $K = 10$

(d) Normalized Throughput U with Different Number of Channel K via Different Probability of Transmission Attempt p, $N_{cw} = 5$, $K = 10$

Fig. 8. Validation

It should be noted that Eq.(23)can be solved same as p^* over convex domains. Eq.(23) provides a basic understanding about the optimal performance of channel selection algorithms. In the following section, we simulate different channel selection algorithms to provide some intuition of different selection algorithms.

7. Simulation results

7.1 Validation

First we validate and discuss the protocol performance. Fig. 8 illustrates the analytic and the simulation results of CR multichannel MAC. The simulation results (depicted in dots) validate our analytic model, providing a basic understanding about CR multichannel MAC performance. Note that in this part we turn off the cognitive functionality to analyse the original protocol performance. The results also suggest the correctness of the steady state approximation, which provides important insights for the protocol design and further optimization processes.

Fig. 8-(a) depicts the normalized throughput via CR attempt rate G. The simulation parameters are with $K = 10, 20, 40$; $\Gamma = \{\frac{1}{K}\}$, $\eta = 0.95$, $N_{cw} = 5$, $q_r = 0.2$, $q_k = 0.05$. One can observe that with lower number of channel (resource), the MAC throughput will saturate faster (under lower attempt rate G), and then drop faster. The more the number of channel, the higher attempt rate the CR can support. It matches the intuition that with more resource, more CRs can be served. Without a proper stabilization mechanism, the normalized throughput under all λ drops as attempt rate G increased.

With different number of channels, different attempt rates of CR can be supported. Fig. 8-(b) illustrates the normalized MAC throughput via the number of channels. With a given transmission attempt G, when the number of channel is large, there are less contention when CRs selects over multichannel. On the other hand, the appearance of PSs' users will affect CR's multichannel MAC performance. In Fig. 8-(c), the higher the probability of PS appearance, the lower the CR multichannel MAC throughput will be achieved. Note that with the spectrum sensing, transmissions of PSs' users are protected.

The result in Fig. 8-(d) shows three cases when the number of CR is much greater than the number of channels, that is, a crowded CRN; about the number of channels and less then the number of channels. It can be concluded that when the number of CR is less then the channel, it is reasonable to transmit it with higher probability when packet arrives. When the number of CR is higher, it should decrease the probability of transmission attempt q to alleviate the contention over multichannel. The dramatic drop in throughput is more severe in a crowded CRN.

7.2 Cognitive multichannel MAC performance

For the proposed cognitive CSMA-based multichannel MAC protocol, the throughput outperforms the the conventional work of Aloha-based McMAC (Mo et al., 2008) with improvement in PS co-existence support and collision alleviation of CRs by CSMA. Fig. 9 shows the normalized/aggregated throughput via number of CRs and channels. (a) and (b) is the McMAC case while (c) and (d) is the cognitive multichannel MAC protocol. With the joint consideration of channel selection and contention avoidance/resolution, more sophisticated CR multichannel MAC and better normalized/aggregated throughput performance can be achieved. Note that the cognitive functionality is turned off for a fair comparison.

For normalized throughput/utilization of the cognitive multichannel MAC and McMAC respectively depicted in Fig. 9-(c) and Fig. 9-(a), co-channel intra-system contention in cognitive multichannel MAC is separated over multichannel and is further alleviated by the non-persistent CSMA, therefore, the throughput is enhanced. When the number of channel is low, this phenomena is significant. when the number of channel increase, the alleviation of co-channel contention is still valid in the improvement of channel utilization. The out performance comes from the smarter channel selection and the contention resolution among PS-CR and CR-CR.

The aggregated throughput performance is illustrated in Fig. 9-(d) and Fig. 9-(b). When the number of channel is increased, the throughput is first increased and then converged. It is because that p and N are fixed, CRs can not initiate more parallel transmissions even if there

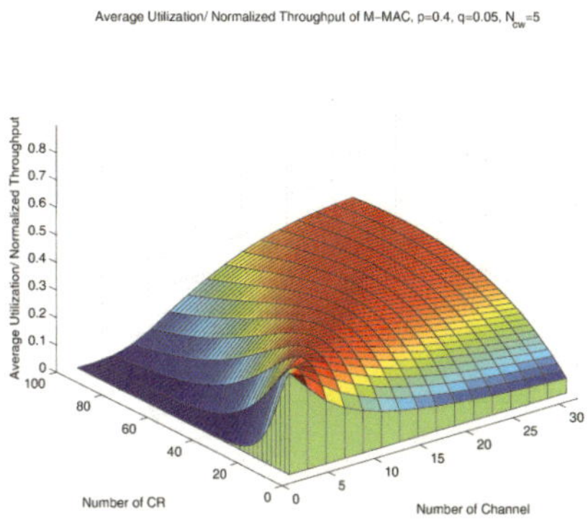

(a) Normalized Throughput/ Average Utilization of McMAC

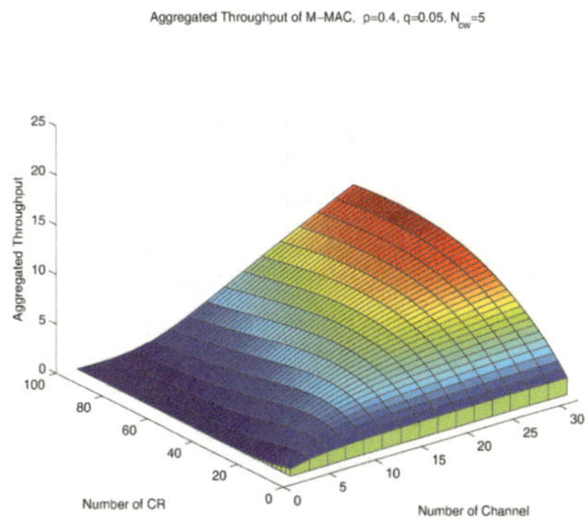

(b) Aggregated Throughput of McMAC

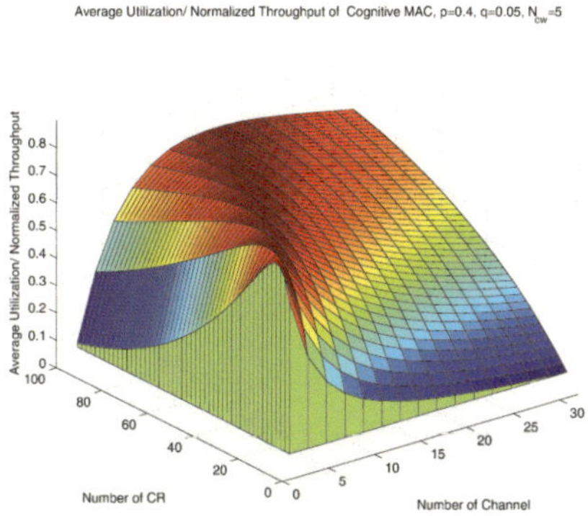

(c) Normalized Throughput/ Average Utilization of Cognitive MAC

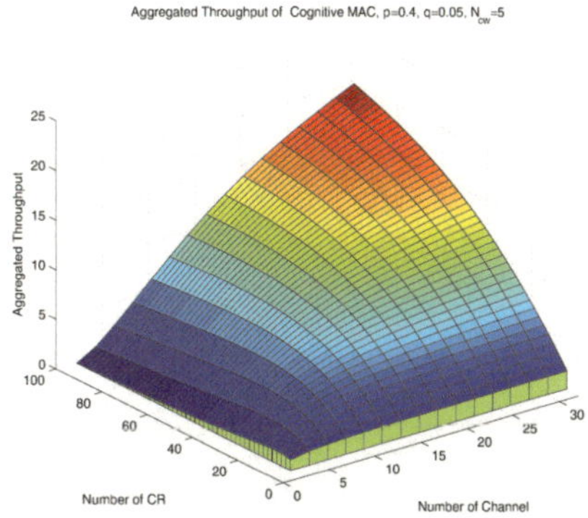

(d) Aggregated Throughput of Cognitive MAC

Fig. 9. Multichannel MAC performance

is still available resource. On the other hand, for a fix number of channels, increasing the size of CRN results in the increase of throughput since there are more CRs try to transmit, which is a feasible property of CR multichannel MAC.

7.3 Optimal cognitive multichannel MAC

With the cognizant information in Section 6, the proposed cognitive multichannel MAC can further been optimized. Fig. 10-(a) shows the normalized throughput with stabilization that maximizes throughput with $p*$. In addition to spectrum sensing and CSMA, with the help of cognizant information from PSs and CRN, the proposed cognitive multichannel protocol empowers CRN to adapt to the optimal probability of transmission attempt p^* for pseudo-Bayesain algorithm. Without losing intuition, here we assume CRN without entering/leaveing CRs, i.e., no transient behaviour. In our simulation, we consider the case that the environmental information including the number of CR N, the probability of PS appearance q_k is correctly and distributively acquired by all CRs in the long run. The channel capacity C is set to unity and the size of contention window $N_{cw} = 5$. The probability of PS

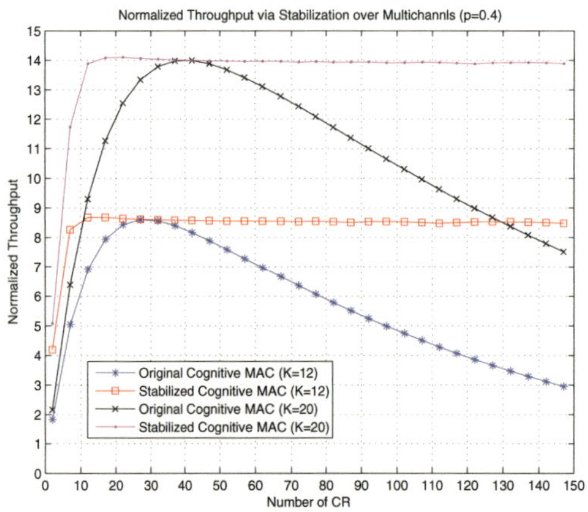

(a) Protocol Stabilization

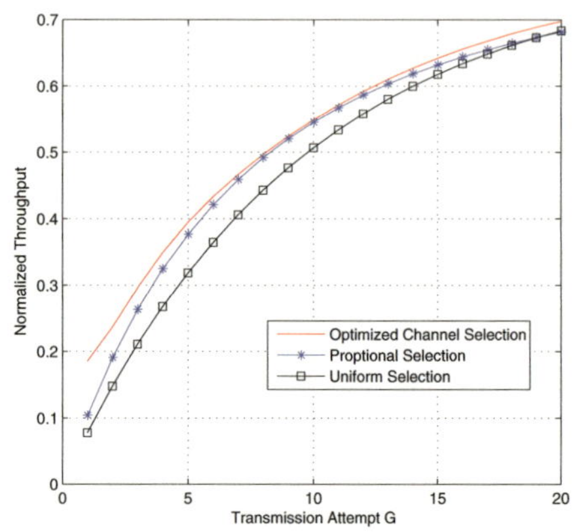

(b) Optimal Channel Selection

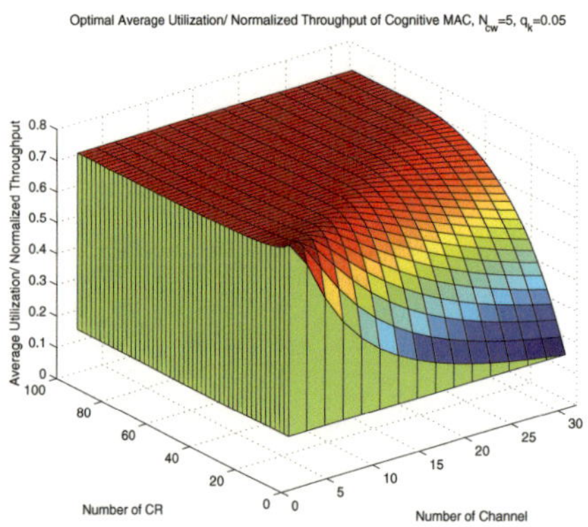

(c) Optimal Normalized Throughput/ Average Utilization of Cognitive MAC

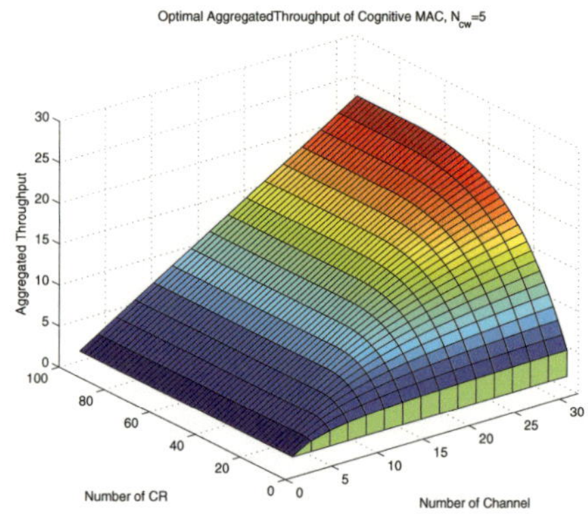

(d) Optimized Aggregated Throughput of Cognitive MAC

Fig. 10. MAC Protocol Optimization with Cognitive Functionality

appearance is 0.05. We simulate the case $K = 12$ and $K = 20$ as a representation of stabilization of the proposed cognitive multichannel MAC protocol.

Observing Fig. 10-(b), the optimal channel selection outperforms all other strategies. When the number of CR selections N is low, its performance is equivalent to best selection, that is, the optimal selection algorithm selects the networks offering maximum utility. It should be noted that when N increases, always selecting the best channel is no more the 'best' strategy. It is due to the fact that best selection saturates the best channels, leaving other channels far away from fully utilized.

Opportunistic access including random and proportional selection may be used to achieve expected utility improvement and load balancing. From the simulation results, proportional selection outperforms random selection the most of the time. When number of CR N is not large, selecting best channels is a feasible manner, however, when N is increasing, proportional channel selection is close to the optimal selection strategy. When the number of

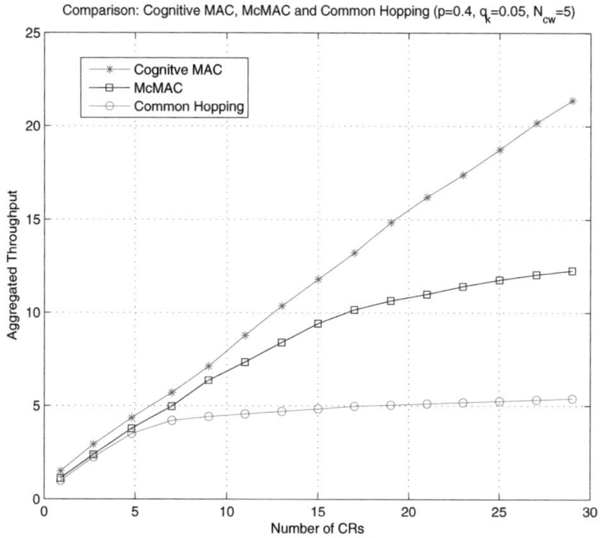

Fig. 11. Comparison of the proposed Cognitive MAC and other multichannel MAC

CR N is high, random selection may be a good asymptotic channel selection strategy. While the number of CR N is low, proportional selection will be a feasible approach.

Fig. 10-(c) and (d) depict normalized and aggregated throughput R of the cognitive multichannel MAC with the cognitive functionalities for MAC protocol optimization. From the result, we further show that adaptation of p^* and Γ^* will optimize the throughput performance of multichannel MAC. The aggregated MAC throughput performance can be improved in comparison to the converged phenomena of non-cognitive protocol shown in Fig. 9. The average improvement is 27.4% under our simulation setting. The insight behind is : CR may increase its attempt probability and choose the better channel when system is in light load. While when the system is crowded, CR may adapt and selecting channel base on the optimization rule p^* and Γ^*. MAC protocol optimization rely on the cognitive functionality, which will be a promising capability in future communication system such as CRN. From above, the system is stabilized by adapting p^*, therefore, the overall throughput is approaching the upper bound under all condition of CRN. Moreover, the upper bound is increased by optimal channel selection strategy Γ^* of CR. This optimization further improves the original out-performance of the proposed cognitive multichannel MAC for CR.

In Fig. 11, the proposed cognitive multichannel MAC protocol is summarized out-performance over other protocol designs under all size of CR since it can increase p to initiate more transmissions while there is more resource for the CRs and decrease p to avoid contention and collisions while the resource is relatively rare. From above, the cognitive functionality and cognizant information of CR provide further possibility of protocol performance improvement. The speed and accuracy of information retrieval will affect the performance. More sophisticated information extracting/exchanging mechanisms left much room for future research.

8. Conclusion

The future multichannel MAC under CR paradigm is very critical but challenging. CR must efficiently utilize the left spectrum resource while avoid interfering PSs' users. The key components in the multichannel MAC are collision avoidance/resolution and channel selection. In this chapter, a general multichannel MAC protocol scheme, the related mathematical model and the steady state approximation is proposed to build up the insight of CR multichannel MAC problem, solution and evaluation. The cognitive CSMA-based multichannel example suggests throughput enhancement for CR in comparison of current research status.

The cognitive functionality of cognitive radio makes future CR multichannel MAC more intelligent. We further develop and integrate the idea of *cognition* into MAC design of CRN. The proposed cognitive multichannel MAC protocol utilizes CRs' *sensing* and *adaption* functionality to extract both inter- and intra-system information and optimize the throughput performance. For both PSs' and CRs' operators, the proposed cognitive multichannel MAC protocol is feasible since it offers detection and avoidance for PSs, and provides diverse multichannel accessibility and adaptation that optimizes the MAC throughput performance. Simulation results further validate the cognitive functionality and show its significance. Future research in MAC for CR or cognition functionality are obviously crucial and interesting, and the cognitive multichannel MAC protocol for CRN is an exciting first step.

9. References

Akyildiz, I. F., Lee, W.-Y., Vuran, M. C. & Mohanty, S. (2006). Next generation/dynamic spectrum access/cognitive radio wireless networks: a survey, *Comput. Networks* 50(13): 2127–2159.
URL: *http://dx.doi.org/10.1016/j.comnet.2006.05.001*

Al-Meshhadany, T. & Ajib, W. (2007). New cdma-based mac protocol for ad hoc networks, *Vehicular Technology Conference, 2007. VTC-2007 Fall. 2007 IEEE 66th* pp. 91–95.

Bahl, P., Chandra, R. & Dunagan, J. (2004). Ssch: slotted seeded channel hopping for capacity improvement in ieee 802.11 ad-hoc wireless networks, *MobiCom '04: Proceedings of the 10th annual international conference on Mobile computing and networking*, ACM, New York, NY, USA, pp. 216–230.

Bertsekas, D. & Gallager, R. (1992). *Data Networks*, second edn, Prentice Hall.

Boyd, S. P. & Vandenberghe, L. (2004). *Convex Optimization*, Cambridge University Press.

Chen, J., Sheu, S.-T. & Yang, C.-A. (2003). A new multichannel access protocol for ieee 802.11 ad hoc wireless lans, *Personal, Indoor and Mobile Radio Communications, 2003. PIMRC 2003. 14th IEEE Proceedings on* 3: 2291–2296 vol.3.

Chen, K.-C., Peng, Y.-J., Prasad, N., Liang, Y.-C. & Sun, S. (2008). Cognitive radio network architecture: part i – general structure, *ICUIMC '08: Proceedings of the 2nd international conference on Ubiquitous information management and communication*, ACM, New York, NY, USA, pp. 114–119.

Cordeiro, C. & Challapali, K. (2007). C-mac: A cognitive mac protocol for multi-channel wireless networks, *New Frontiers in Dynamic Spectrum Access Networks, 2007. DySPAN 2007. 2nd IEEE International Symposium on* pp. 147–157.

Haykin, S. (2005). Cognitive radio: brain-empowered wireless communications, *Selected Areas in Communications, IEEE Journal on* 23(2): 201–220.
URL: *http://dx.doi.org/10.1109/JSAC.2004.839380*

IEEE (2009). IEEE Standard Committee 802.22 Working Group on Wireless Regional Area Networks, *IEEE Std 802.22-2009* .

Jain, N., Das, S. & Nasipuri, A. (2001). A multichannel csma mac protocol with receiver-based channel selection for multihop wireless networks, *Computer Communications and Networks, 2001. Proceedings. Tenth International Conference on* pp. 432–439.

Jakllari, G., Luo, W. & Krishnamurthy, S. (2005). An integrated neighbor discovery and mac protocol for ad hoc networks using directional antennas, *World of Wireless Mobile and Multimedia Networks, 2005. WoWMoM 2005. Sixth IEEE International Symposium on a* pp. 11–21.

Jha, S., Rashid, M., Bhargava, V. & Despins, C. (2011). Medium access control in distributed cognitive radio networks, *Wireless Communications, IEEE* 18(4): 41 –51.

Limb, K. K. J. O. (1987). *Advances in Local Area Networks. Frontiers in Communications*, IEEE Press.

Liu, X. & Ding, Z. (2007). Escape: A channel evacuation protocol for spectrum-agile networks, *New Frontiers in Dynamic Spectrum Access Networks, 2007. DySPAN 2007. 2nd IEEE International Symposium on* pp. 292–302.

Marsan, M. & Roffinella, D. (1983). Multichannel local area network protocols, *Selected Areas in Communications, IEEE Journal on* 1(5): 885–897.

Mitola, J., I. & Maguire, G.Q., J. (1999). Cognitive radio: making software radios more personal, *Personal Communications, IEEE* 6(4): 13–18.

Mo, J., So, H.-S. & Walrand, J. (2008). Comparison of multichannel mac protocols, *Mobile Computing, IEEE Transactions on* 7(1): 50–65.

Nasipuri, A., Zhuang, J. & Das, S. (1999). A multichannel csma mac protocol for multihop wireless networks, *Wireless Communications and Networking Conference, 1999. WCNC. 1999 IEEE* pp. 1402–1406 vol.3.

NTIA (2003). U.s. frequency allocation.
URL: *http://www.nitia.doc.gov/osmhome/allochrt.pdf*

Report, F. & Order (2002). Federal communication commission std., *FCC* 02: 48.

RoyChoudhury, R., Bandyopadhyay, S. & Paul, K. (2000). A distributed mechanism for topology discovery in ad hoc wireless networks using mobile agents, *MobiHoc '00: Proceedings of the 1st ACM international symposium on Mobile ad hoc networking & computing*, IEEE Press, Piscataway, NJ, USA, pp. 145–146.

So, J. & Vaidya, N. H. (2004). Multi-channel mac for ad hoc networks: handling multi-channel hidden terminals using a single transceiver, *MobiHoc '04: Proceedings of the 5th ACM international symposium on Mobile ad hoc networking and computing*, ACM, New York, NY, USA, pp. 222–233.

Sun, Y.-K., Chen, K.-C. & Twu, D.-C. (1997). Generalized tree multiple access protocols in packet switching networks, *Personal, Indoor and Mobile Radio Communications, 1997. 'Waves of the Year 2000'. PIMRC '97., The 8th IEEE International Symposium on* 3: 918–922 vol.3.

Tang, Z. & Garcia-Luna-Aceves, J. (1998). Hop reservation multiple access (hrma) for multichannel packet radio networks, *Computer Communications and Networks, 1998. Proceedings. 7th International Conference on* pp. 388–395.

Timmers, M., Dejonghe, A., van der Perre, L. & Catthoor, F. (2007). A distributed multichannel mac protocol for cognitive radio networks with primary user recognition, *Cognitive Radio Oriented Wireless Networks and Communications, 2007. CrownCom 2007. 2nd International Conference on* pp. 216–223.

Tzamaloukas, A. & Garcia-Luna-Aceves, J. (2000). Channel-hopping multiple access, *Communications, 2000. ICC 2000. 2000 IEEE International Conference on* 1: 415–419 vol.1.

WC Hung, KLE Law, A. L.-G. (2002). A dynamic multi-channel mac for ad hoc lan, *21st Biennial Symposium on Communications* .

Wu, S.-L., Lin, C.-Y., Tseng, Y.-C. & Sheu, J.-L. (2000). A new multi-channel mac protocol with on-demand channel assignment for multi-hop mobile ad hoc networks, *Parallel Architectures, Algorithms and Networks, 2000. I-SPAN 2000. Proceedings. International Symposium on* pp. 232–237.

Zhang, L., Soong, B.-H. & Xiao, W. (2003). A new multichannel mac protocol for ad hoc networks based on two-phase coding with power control (tpcpc), *Information, Communications and Signal Processing, 2003 and the Fourth Pacific Rim Conference on Multimedia. Proceedings of the 2003 Joint Conference of the Fourth International Conference on* 2: 1091–1095 vol.2.

Zhang, X. & Su, H. (2011). Cream-mac: Cognitive radio-enabled multi-channel mac protocol over dynamic spectrum access networks, *Selected Topics in Signal Processing, IEEE Journal of* 5(1): 110 –123.

4

Delay Analysis and Channel Selection in Single-Hop Cognitive Radio Networks for Delay Sensitive Applications

Behrouz Jashni
Iran

1. Introduction

Scarcity and the value of spectrum resource for wireless communications challenged the traditional static spectrum allocation policy. Recent measurements [1] show that in spite of the fact that available frequency spectrum have been almost occupied by the existing carriers, most of the time we can find spectrum bandwidths that are free of signal. Thus, the efficient usage of limited spectrum has been highly required.

Cognitive Radio (CR) technology is an innovative radio design technology that is proposed to increase the spectrum utilization by exploiting the unused spectrum in dynamically changing environments.

With ever-increasing demand for delay sensitive applications such as Video telephony, live audio, surveillance, live video and etc, special attention to this field seems to be essential. In such applications the receiver needs to get transmitted information within a certain delay. Therefore delay study in CR Networks (CRNs) and introducing appropriate channel selection strategy that reduces end-to-end packet transmission delay is critical.

There are two main challenges in the CRNs [2]: a) how to sense the spectrum and model the behavior of the primary licensees to identify available frequency channels (spectrum holes) b) Management of available spectrum resources among Secondary Users (SUs) to satisfy their Quality of service (Qos) requirements while limiting the interference to the primary licensees. In this chapter, we focus on the spectrum management, specially delay analysis and relay on the existing literatures for the first challenge [3]- [5].

In the most of resource management researches in CRNs, the focus is on the single-hop wireless infrastructure. Prior studies such as [6], [7] presented a centralized solution; However due to the informationally decentralized nature of the wireless networks information, the complexity of the optimal centralized solution for spectrum allocation is prohibitive [8] for delay sensitive applications. Moreover, in the centralized solutions, propagation of private information back and forth to a common coordinator causes delay and may be unacceptable for delay sensitive applications. Decentralized solutions are presented in [9]- [11]. In these researches a utility function proposed and with use of several solutions (game theory, statistical methods and etc) try to reach optimum situation, but they do not pay attention to packet delay induced by the network and not suitable for real time applications. Authors in [12] presented an approximation of the end-to-end delay of packets in a single-hop CRN. They proposed a distributed channel allocation algorithm based on iterative delay minimization and utility improvement. They considered the centralized topology with preemptive resume policy and derived an approximate solution. In [13] delay calculations are done for an *M/D/1* priority queueing scheme. They study time slotted CR system with one primary and one secondary user. A priority-concerned MAC protocol based on queue theory was proposed for cognitive radio network in [14]. The priority-concerned optimal MAC scheme following the stopping rule is validated by network satisfaction simulation. In [15], authors study the effect of peak interference power constraint on outage probability and transmission time of packets for SU in a spectrum sharing environment. They consider only two SU (one transmitter (Tx) and one receiver (Rx)) and one PU (Rx). Simultaneously transmission policy are used, that means SU-Tx must keep its power below a certain level such that it does not cause harmful interference to PU-Rx. Authors develop an opportunistic spectrum scheduling scheme for cognitive radio networks in [16]. Each secondary user based on the queue size and the observed channel conditions, estimate the throughput for each channel. A scheduling algorithm is performed to maximize the expected aggregate throughput of all the secondary users. They only consider the time slotted scenario with a centralized decision making system. A simplified model of primary user interruptions (Markov chain) is proposed in [17], queuing analysis is carried out for two-server-single-queue (a single secondary user and two licensed channels) and single-server-two-queue (two secondary users and single channel) cases. A semi-analytic result is obtained for the generating function of queue length in the two-server-single-queue case. The average queue length is obtained from solving a group of linear equations for the single-server-two-queue case.

In this chapter, we consider different scenarios such as centralized and decentralized spectrum management. Introduced new delay analysis and preemtive repeat, preemptive resume and time slotted regimes are investigated. We propose new scheme and formulation for delay calculation in single-hop cognitive radio networks with multiple primary and secondary users. Simulation examples are employed to evaluate the performance of our derivations, showing more accurate and less complicated results comparing with the results in [12]. A prepare channel selection strategy based on channels condition and queues situation are proposed.

The remaining of this chapter is as follows. In Section 2 assumptions and system properties are presented. Delay analysis in single-hop networks for different transmission policies are studied in Section 3. Section 4 states the channel selection strategy. In Section 5, simulation results are presented and Section 6 concludes the chapter.

2. Assumptions and system specifications

Assume that we have N secondary users and M primary users/frequencies in a CRN. All nodes are in the transmission range of each other and there is a link between each two nodes. We must note that if two users transmit data simultaneously on a specific frequency band, we have a collision and damaged data retransmission is required according to IEEE 802.11/e [18].

Based on the spectrum sensing results, vacant frequency bands are determined. Then in order to allocate available frequency channels among the SUs, an appropriate channel allocation scheme that provide their required Qos while not causing any trouble for PUs is needed. The PUs have priority over the SUs and there is no need to be worry about SUs transmission. A preemptive priority queueing model is developed to describe the traffic behavior of users in CRNs; PUs are the owners of the spectrum, while the SUs may use that when they are unoccupied. Here we employ *M/G/1* queueing model, in which the packet entry of users follows the poisson process, the transmission time has general distribution and there is one server in each queue. Packets in each node select an appropriate frequency based on the transmission strategy and join to the corresponding node/frequency queue.

3. Delay analysis

3.1 Preemptive resume policy

Primary users are the padrone of frequencies and are able to preempt the transmission of secondary users. Therefore, cognitive users must release the frequency channel as soon as possible if PU intends to use it. In the preemptive resume regime, SU resumes its interrupted transmission after disconnecting PU.

In this case, delays that each packet experienced is divided into three sections and are shown in Fig.1.

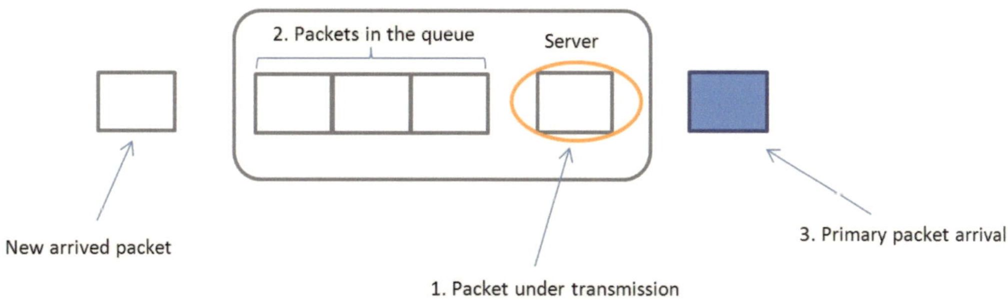

Fig. 1. Waiting delays in the queue of Secondary node

3.1.1 Delay due to waiting for completion of packet servicing under transmission

When a packet enters the queue of a node/frequency, may be another packet under servicing. Thus new packet must wait for completion of that servicing. The mean residual life of packet servicing in *M/G/1* queue calculated as follows [19]: $\frac{E[S^2]}{2E[S]}$ that $E[S^2]$ and $E[S]$ are second and first moment of service time and computed as follows at node *i* on frequency *j:*

$$E[S_{ij}] = \frac{L + L_{oh}}{R_{ij}(1 - P_{ij}^{e})}. \tag{1}$$

$$E[S_{ij}^2] = \frac{(L + L_{oh})^2 (1 + P_{ij}^e)}{R_{ij}^2 (1 - P_{ij}^e)^2} . \tag{2}$$

Here L is the average packet length and L_{oh} is the control header including the time for protocol acknowledgment, information exchange, channel sensing and etc [18]. R_{ij} is the transmission rate and P_{ij}^e is the packet sending error probability that depend on channel conditions [20]. The first and second moment of traffic load (also known as utilization factor) for user i on frequency channel j is [12], [21]:

$$\rho_{ij} = \lambda_{ij} E[S_{ij}] . \tag{3}$$

$$\rho_{ij}^2 = \lambda_{ij} E[S_{ij}^2] . \tag{4}$$

That λ_{ij} is the packet arrival rate at node i on frequency j.

Since ρ_{kj} represents the utilization factor of the service, the conditional probability, that user k's packet is being transmitted on frequency j given that no primary packet is transmitted, is $\frac{\rho_{kj}}{1 - \rho_{0j}}$ (subtitle 0 is used for PU in each frequency), As for primary user the probability of utilization is simply ρ_{ij} [21]. These allow us to formulate first part of waiting delay ($E[W_{ij}]_1$) as follows:

$$\begin{aligned} E[W_{ij}]_1 &= \rho_{0j} \frac{E[S_{0j}^2]}{2E[S_{0j}]} + \sum_{k=1}^{N} \frac{\rho_{kj}}{1 - \rho_{0j}} \frac{E[S_{kj}^2]}{2E[S_{kj}]} = \ldots \\ &= \frac{\rho_{0j}^2}{2} + \frac{1}{1 - \rho_{0j}} \sum_{k=1}^{N} \frac{\rho_{kj}^2}{2} \end{aligned} \tag{5}$$

3.1.2 Delay due to waiting for service of the packets exist in the queue

Since there is a probability that some packets in the queue, new packet must waits for servicing these older packets. Number of existed packets in the queue of node i on frequency j based on little's theorem [19], [22] is:

$$E[No_{ij}] = \lambda_{ij} E[W_{ij}] . \tag{6}$$

In each moment, only one user can transmit data on specified frequency. Thus each packet on the queue of frequency j at node i also wait for servicing the older packets of other nodes. Number of all packets on frequency j is:

$$\sum_{k=0}^{N} E[No_{kj}] . \tag{7}$$

These packets servicing delay are computed as follows:

$$E[W_{ij}]_2 = \sum_{k=0}^{N} E[No_{kj}]E[S_{kj}] = \sum_{k=0}^{N} \rho_{kj}E[W_{kj}]. \quad (8)$$

3.1.3 Delay due to new primary packets arrival

When a packet wait for servicing, it's probably that new primary packets enter to system and because of upper priority of these packets, the servicing time of them added to waiting time of secondary packets in the queue.

Number of new primary packets entry $E[No'_{ij}] = \lambda_{0j}E[W_{ij}]$ and required time to service of these packets are equal:

$$E[W_{ij}]_3 = E[No'_{ij}]E[S_{ij}] = \rho_{0j}E[W_{ij}]. \quad (9)$$

Using these three delays, we can calculate the overall waiting time of secondary users in the queue of node *i* and frequency *j* as follows:

$$E[W_{ij}] = E[W_{ij}]_1 + E[W_{ij}]_2 + E[W_{ij}]_3 = \ldots$$

$$= \frac{\frac{\rho_{0j}}{2} + \frac{1}{1-\rho_{0j}} \sum_{k=1}^{N} \frac{\rho_{kj}^2}{2} + \rho_{0j}E[W_{0j}]}{1 - \sum_{k=0}^{N} \rho_{kj}}. \quad (10)$$

For primary users, we can obtain the mean waiting time of a packet in a similar method. Note that we do not have $E[W_{0j}]_3$ because of primary users will not be intercepted by any users.

$$E[W_{0j}] = E[W_{0j}]_1 + E[W_{0j}]_2 = \frac{\rho_{0j}^2}{2(1-\rho_{0j})}. \quad (11)$$

Finally, the average waiting delay in preemptive resume policy is:

$$E[W_{ij}] = \frac{\sum_{k=0}^{N} \rho_{kj}^2}{2(1-\rho_{0j})(1-\sum_{k=0}^{N} \rho_{kj})}. \quad (12)$$

3.2 Preemptive resume policy

Since in the wireless environment, transmission of packet resumption is impossible. We practically can not use from preemptive resume policy. In such cases, after completion of primary transmission, secondary user, retransmit its packet. In this method the waiting delay calculation is as follows:

3.2.1 Delay of waiting for completion of packet under transmission

This delay is depending on the probability of PUs or another SUs packets presentation and equal:

$$E[W_{ij}]_1 = \frac{\rho_{0j}}{2} + \frac{1}{1-\rho_{0j}} \sum_{k=1}^{N} \frac{\rho_{kj}^2}{2}. \tag{13}$$

3.2.2 Delay due to waiting for service of packets exist in the queue

New arrived packet should waits for servicing of all packets (PU and SUs) are queued on specified frequency and computed as follows:

$$E[W_{ij}]_2 = \sum_{k=0}^{N} \rho_{kj} E[W_{kj}]. \tag{14}$$

3.2.3 Delay due to new primary packets arrival

This waiting time is equal probability of PUs new packet arrival multiply mean waiting time and calculated like this:

$$E[W_{ij}]_3 = \rho_{0j} E[W_{ij}]. \tag{15}$$

3.2.4 Delay due to incomplete servicing of some secondary packets

When a new primary packet arrives, occupy the frequency channel and SU that sending packet must stop its transmission and after ending occupancy of frequency channel by the PU, retransmit its packet. It's probable that transmission of some secondary packets stopped by primary users. Since utilization factor of node i on frequency j is $\frac{\rho_{kj}}{1-\rho_{0j}}$, then number of primary transmission that collided by transmission of node i on frequency j is:

$$\lambda_{0j} E[W_{ij}] \frac{\rho_{ij}}{1-\rho_{0j}}. \tag{16}$$

Time elapsed of servicing each packet is defined as "Age" and for *M/G/1* queue like residual life calculated [15]. Now we can calculate this incomplete servicing computed as follows:

$$\begin{aligned} E[W_{ij}]_4 &= \sum_{k=1}^{N} \lambda_{0j} E[W_{kj}] \frac{\rho_{kj}}{1-\rho_{0j}} \frac{E[S_{kj}^2]}{2E[S_{kj}]} = \ldots \\ &= \frac{\lambda_{0j}}{1-\rho_{0j}} E[W_{ij}] \sum_{k=1}^{N} \rho_{kj}^2 \end{aligned} \tag{17}$$

The overall waiting delay in preemptive repeat regime is calculated like this:

$$E[W_{ij}] = E[W_{ij}]_1 + E[W_{ij}]_2 + E[W_{ij}]_3 + E[W_{ij}]_4$$
$$= \frac{\frac{\rho_{0j}^2}{2} + \frac{1}{1-\rho_{0j}} \sum_{k=1}^{N} \frac{\rho_{kj}^2}{2} + \rho_{0j} E[W_{0j}]}{1 - (\rho_{0j} + \sum_{k=0}^{N} \rho_{kj} + \frac{\lambda_{0j}}{1-\rho_{0j}} \sum_{k=1}^{N} \rho_{kj}^2)} \quad (18)$$

By substituting $E[W_{0j}]$ from (11) we have:

$$E[W_{ij}] = \frac{\sum_{k=0}^{N} \rho_{kj}}{2(1-\rho_{0j})[1 - (\rho_{0j} + \sum_{k=0}^{N} \rho_{kj} + \frac{\lambda_{0j}}{1-\rho_{0j}} \sum_{k=1}^{N} \rho_{kj}^2)]} \quad (19)$$

3.3 Time slotted and randomizes policy

Here we present time slotted system that packet transmission is done in one time slot. At the beginning of each slot, spectrum sensing is done and vacant frequencies channels determined. Then based on mechanism like P-Slotted-Alloha [22], packets are transmitted. Each packet waits random slots that equal random number with uniform distribution from *[0,w]* and transmit the packet. In this case, we do not have control on packet transmission on several frequencies, so there is a probability of packet collision. If collision occurred, random waiting and retransmission is done.

In this system, the probability of packet transmission on a slot is:

$$P_{ij}^t = \rho_{ij} \frac{2}{\omega + 1} . \quad (20)$$

We have success transmission on a slot if only one user forward packet on that slot and its probability is:

$$P_{ij}^{succ} = \rho_{ij}^t \prod_{k=1,k\neq i}^{N} (1 - P_{kj}^t) . \quad (21)$$

The probability of error in packet transmission is:

$$P_{ij}^e = P_{ij}^t - P_{ij}^{succ} . \quad (22)$$

Now we can calculate the first and second moment of the service time as follows, that T is the time slot length.

$$E[S_{ij}] = \frac{T(1+\frac{\omega}{2})}{1 - P_{ij}^e} . \quad (23)$$

$$E[S_{ij}^2] = \frac{T^2(1+\frac{\omega}{2})(1+P_{ij}^e)}{(1-P_{ij}^e)^2} \tag{24}$$

The waiting delay in such system computed as follows:

3.3.1 Delay of waiting for service completion of packet under transmission

That composed of two components: probability of PU's packet under transmission or SU's packet.

$$E[W_{ij}]_1 = \rho_{0j} \frac{E[S_{0j}^2]}{2E[S_{0j}]} + \frac{\rho_{ij}}{1-\rho_{0j}} \frac{E[S_{ij}^2]}{2E[S_{ij}]}. \tag{25}$$

3.3.2 Delay due to primary packets arrivals

Because of, transmission in each node independent of other nodes, new packet in the queue of specified frequency only waits for servicing of these queued packets.

$$E[W_{ij}]_2 = (\lambda_{ij} E[W_{ij}])E[S_{ij}]. \tag{26}$$

3.3.3 Delay due to waiting for service of packets exist in the queue

When a packet transmission is cancelled by PUs occupancy, random waiting time and retrains mission is done.

$$E[W_{ij}]_3 = \lambda_{0j} E[W_{0j}](E[W_{0j}] + E[W_{ij}]). \tag{27}$$

Finally the overall waiting delay is equal:

$$E[W_{ij}] = E[W_{ij}]_1 + E[W_{ij}]_2 + E[W_{ij}]_3 = \dots$$
$$= \frac{\rho_{0j} \frac{E[S_{0j}^2]}{2E[S_{0j}]} + \frac{\rho_{ij}}{1-\rho_{0j}} \frac{E[S_{ij}^2]}{2E[S_{ij}]} + \rho_{0j} E[S_{0j}] \cdot}{(1-\rho_{0j}-\rho_{ij})} \tag{28}$$

If the primary system is time slotted too, $E[W_{0j}]$ calculated as follows:

$$E[W_{0j}] = \frac{E[S^2]}{2E[S]} + \rho_{0j} E[W_{0j}] = \frac{T}{2} + \lambda_{0j} T E[W_{0j}]. \tag{29}$$

So, we have:

$$E[W_{ij}] = \frac{\rho_{0j} T(2-\rho_{0j}) + \rho_{ij} \frac{E[S_{ij}^2]}{2E[S_{ij}]}}{2(1-\rho_{0j})(1-\rho_{0j}-\rho_{kj})}. \tag{30}$$

4. Channel selection and packet transmission process

Here we explain the channel selection and packet transmission process occur on the queues of the network.

- Selection of frequency queue based on transmission strategy. When a packet enters the node, selects a frequency queue with minimum average sojourn (waiting + servicing) time. Thus, probability of selecting frequency j at node i is:

$$\beta_{ij} = \frac{[\sum_{m=1}^{M}(E[W_{im}]+E[S_{im}])^{-1}]^{-1}}{E[W_{ij}]+E[S_{ij}]} . \tag{31}$$

- Determine packet to transmit.
- Send Request To Send (RTS) and wait for response of other side.
- Receive Clear To Send (CTS) and send the packet.
- If the transmission is successful, update the delay vector on several frequencies.
- Else back to step 1 and retransmit the packet.

5. Simulation results

In this section we consider a single-hop CRN with $N=6$ secondary user and $M=3$ primary user/frequency. Average packet length $L=1000B$ and mean transmission rate $R=1.5$ *Mbps*. Because of we want steady state values; we do some experiments and check them to reach steady state. If the results of done experiments have a little variance we accept the obtained outage (mean of results).

For brevity we show only obtained results of one frequency for each experiment.

5.1 On demand transmission

In the first example we increase the traffic of network (packet arrival rate of users) to 1.5 tantamount. Figures 2, 4 show that increasing of packet arrival rate cause enforcing more delay to packets in the queue at preemptive resume and preemptive repeat regimes respectively. Proximity of simulation results and theory are obvious in the figures.

In the next example, we increase the arrival rate of primary user 3 (on frequency channel 3) and investigate the conversion of frequency selection strategy. We can see in figures 3, 5 that increasing PU3's traffic causes using another frequency channels (1, 2) to transmission of packets (probability of another channel is increased).

5.2 Scheduled system and randomize transmission

In IEEE 802.22 standard that introduce for work in TV broadcasting frequency band, mentioned that 2ms disorder in receiving signal is tolerable [23], Thus we set time slot length $T=2ms$. Packet transmission is done in one slot and number of waiting slot from $\omega \in [0,7]$.

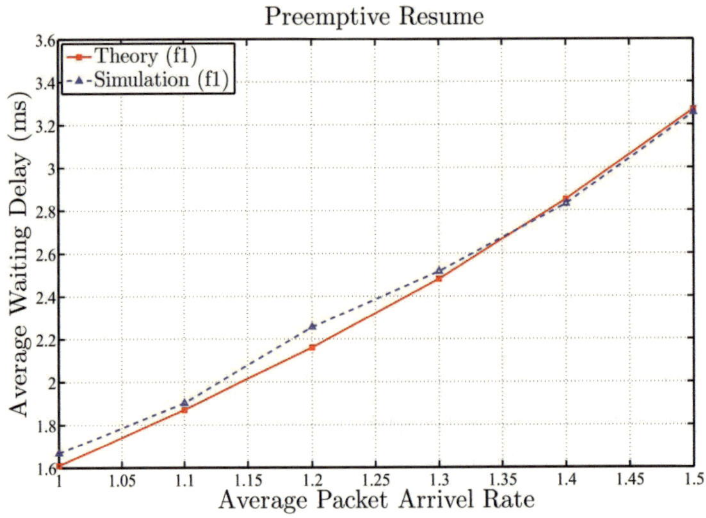

Fig. 2. Comparison of waiting delay formula with simulation results by increasing arrival rates in frequency 1

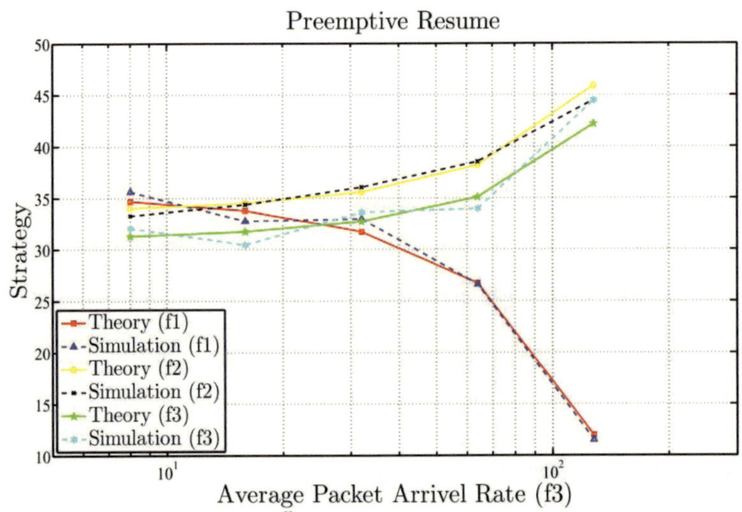

Fig. 3. Impact of arrival rate increasing in frequency 3 on channel selection probability on SU 1

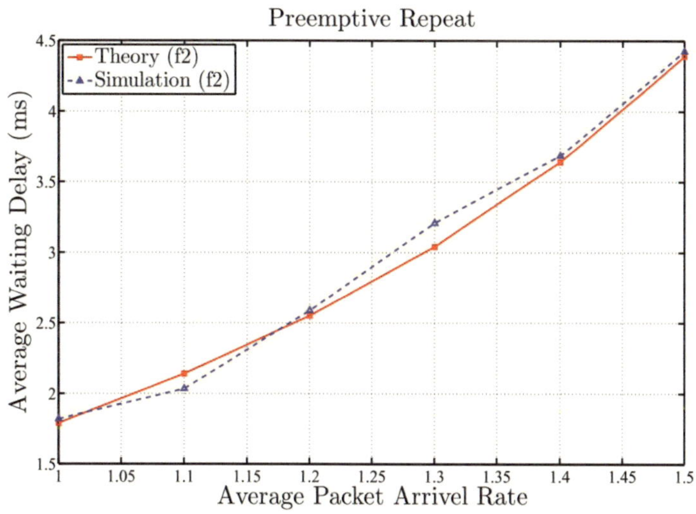

Fig. 4. Comparison of waiting delay formula with simulation results by increasing arrival rates in frequency 2

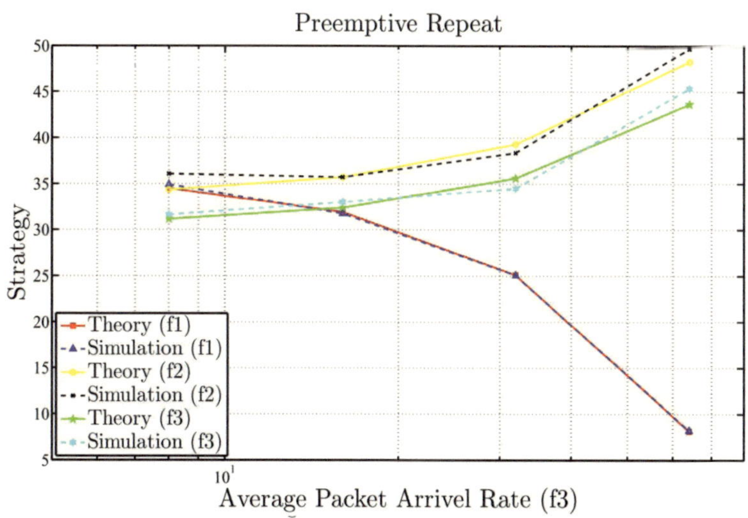

Fig. 5. Impact of arrival rate increasing in frequency 3 on channel selection probability on SU 1

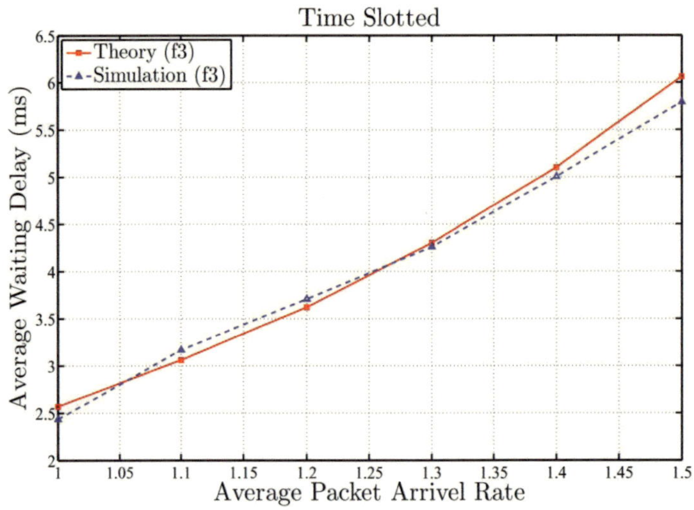

Fig. 6. Comparison of waiting delay formula with simulation results by increasing arrival rates in frequency 3

In figure 6, impact of traffic increasing on packets delay on frequency3 is investigated.

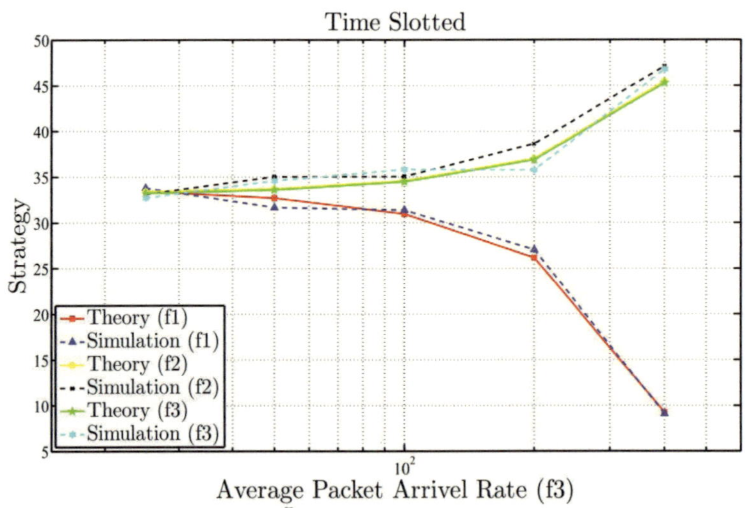

Fig. 7. Impact of arrival rate increasing in frequency 3 on channel selection probability on SU1

Channel selection strategy changing by increasing of PU3's packet arrival rate is shown on figure7.

5.3 Computational complexity

In this section we compare the accuracy of our formulation and introduced formulation in [12]. Results (figures 8-10) show that our calculation is more precise and have less difference with simulation outputs (convergence of new derived formulation is better than the previous one). Complexity comparison is presented in table 1.

Method	+	×
Our formulation	M(2N)	M(4+2N)
Presented formulation in [12]	M(5N+1)+N-1	M(7N+5)

Table 1. Complexity comparison of our introduced formulation with presented formulation in [12]

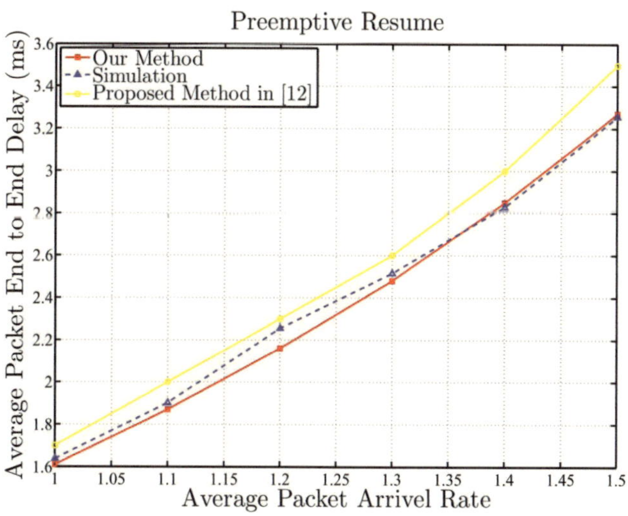

Fig. 8. Comparison of our formulation and presented formulation in [12] (f1)

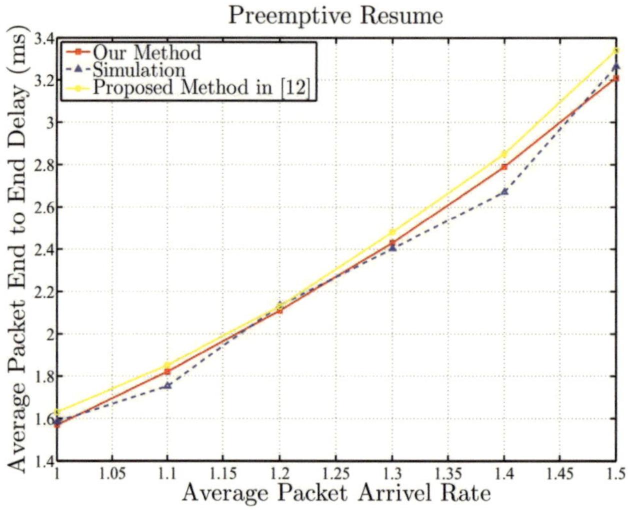

Fig. 9. Comparison of our formulation and presented formulation in [12] (f2)

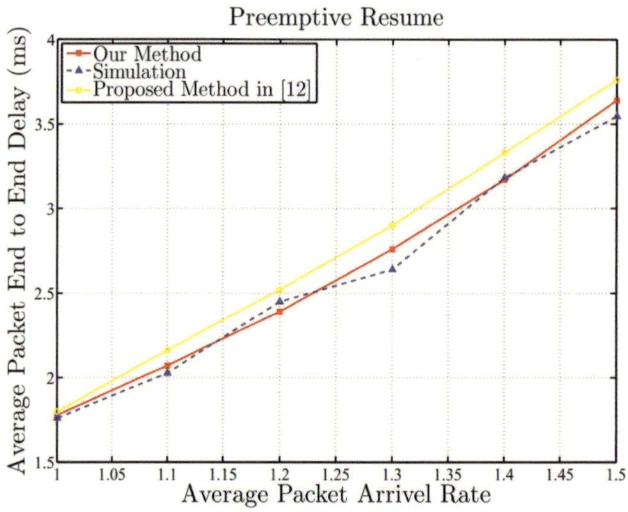

Fig. 10. Comparison of our formulation and presented formulation in [12] (f3)

6. Conclusion

In this chapter, we model the single hop cognitive radio networks and extract the delay of packet transmission with use of queuing theory. Based on history of frequency channels (mean staying time) select an appropriate frequency channel and forward the packet. Here, channel selection and delay analysis in single-hop cognitive radio networks are investigated. Several transmission policies (preemptive resume, preemptive repeat and randomize) are mentioned and studied. Finally with use of simulation investigate the validity and accuracy of obtained terms.

These calculations can be used for delay evaluation in single-hop cognitive radio networks and proposed method suitable for channel selection in delay sensitive applications to provide required Qos. Spectrum management and appropriate link-frequency selection in multi-hop CRNs with emphasis on delay are presented in [24]. In the future work we extend our studies to multi-hop infrastructure and introduce delay analysis of these networks.

7. References

[1] Fed. Commun. Comm. (FCC) (Nov. 2002). *Spectrum Policy Task Force* ,ET Docket no. 02-13515

[2] H. P. Shiang, M. van der Schaar (2009). Distributed Resource Management in Multihop Cognitive Radio Networks for Delay-Sensitive Transmission, *IEEE Transactions on Vehicular Technology*, vol. 58, no. 2, pp. 941-953

[3] I. F. Akyildiz, W.-Y. Lee, M. C. Vuran, S. Mohanty (Sep. 2006). Next generation/dynamic spectrum access/cognitive radio wireless networks: A survey, *Comput. Netw.: Int. J. Comput. Telecommun. Netw.*, vol. 50, no. 13, pp. 2127-2159,

[4] T. X. Brown (2005). An analysis of unlicensed device operation in licensed broadcast service bands, in Proc. *IEEE DySPAN*, pp. 11-29

[5] T. Yucek, H. Arsalan (2009). A survey of spectrum sensing algorithms for cognitive radio applications, *IEEE Communications Surveys & Tutorials*, vol. 11, no. 1, pp. 116-130

[6] C. Cordeiro, K. Challapali, D. Birru, S. Shankar N (Apr. 2006). IEEE 802.22: An introduction to the first wireless standard based on cognitive radios, *J. Commun.*, vol. 1, no. 1

[7] S. A. Zekavat and X. Li (Apr. 2006) Ultimate dynamic spectrum allocation via user-central wireless systems, *J. Commun.*, vol. 1, no. 1, pp. 60-67

[8] C. C. Wang, G. J. Pottie (Oct. 2002). Variable bit allocation for FH-CDMA wireless communication systems, *IEEE Trans. Commun.*, vol. 50, no. 10, pp. 1637-1644

[9] Q. Zhao, L. Tong, A. Swami, and Y. Chen (Apr. 2007). Decentralized cognitive MAC for opportunistic spectrum access in ad hoc networks: A POMDP framework, *IEEE J. Sel. Areas Commun.-Special Issue Adaptive, Spectrum Agile Cognitive Wireless Networks*, vol. 25, no. 3, pp. 589-600

[10] D. Niyato and E. Hossain (Mar. 2007). A game-theoretic approach to competitive spectrum sharing in cognitive radio networks, in Proc. *IEEE WCNC*,Hong Kong, pp. 16-20

[11] Z. Han, C. Pandana, and K. J. R. Liu (2007). Distributive opportunistic spectrum access for cognitive radio using correlated equilibrium and no-regret learning, in Proc. *IEEE Wireless Commun. Netw. Conf.*, pp. 11-15

[12] H. P. Shiang, M. van der Schaar (Aug. 2008). Queueing-based dynamic channel selection for hetrogeneous multimedia applications over cognitive radio networks, *IEEE Trans on Multimedia*, vol. 10, no. 5, pp. 896-909

[13] I. Suliman, J. Lehtomaki (May. 2009). Queueing analysis of opportunistic access in cognitive radios, *Cognitive Radio and Advanced Spactrum Managmanet (CogaART)*, pp. 153-157

[14] X. Zhou, Q. Zhang, H. Zhang, Z. Yan (Oct. 2010). A priority-concerned optimal MAC for cognitive radio networks, *IEEE International Conference on Intelligent Computing and Intelligent Systems (ICIS)*, vol. 3, pp. 314-318

[15] H. Tran, T. Q. Duong, , H.-J. Zepernick (2011). Queuing Analysis For Cognitive Radio Networks Under Peak Interference Power Constraint, *6th International Symposium on Wireless and Pervasive Computing (ISWPC)*, pp. 1-5

[16] V. K. Tumuluru, P. Wang, D. Niyato (2010). An Opportunistic Spectrum Scheduling Scheme for Multi-Channel Cognitive Radio Networks, *72nd IEEE Vehicular Technology Conference Fall (VTC 2010-Fall)*, pp. 1-5

[17] H. Li, Z. Han (2010). Queuing analysis of dynamic spectrum access subject to interruptions from primary users, Proceedings of *the Fifth International Conference on Cognitive Radio Oriented Wireless Networks & Communications (CROWNCOM)*, pp. 1-5

[18] *IEEE 802.11e/D5.0*, Draft Supplement to Part 11: Wireless Medium Access Control (MAC) and physical layer (PHY) specifications: Medium Access Control (MAC) Enhancements for Quality of Service (QoS), (Jun. 2003).

[19] L. Kleinrock (1976). *Queueing Systems Volume 1: Theory*, John Wiley and Sons, New York

[20] D. Krishnaswamy (2002). Network-assisted Link Adaptation with Power Control and Channel Reassignment in Wireless Networks, *3G Wireless Conference*, pp. 165-170

[21] C. Zhang, X. Wang, J. Li (Jun. 2009). Cooperative Cognitive Radio with Priority Queueing Analysis, *IEEE International Conference on Communications, ICC '09*, pp, 1-5

[22] D. Bertsekas and R. Gallager (1987). *Data Networks*, Upper Saddle River, NJ: Prentice-Hall, 1987.

[23] C. Cordeiro, K.Challapali, D. Birru, N. Sai Shankar (Nov. 2005). IEEE 802.22: the first worldwide wireless standard based on cognitive radios, *First IEEE International Symposium on New Frontiers in Dynamic Spectrum Access Networks (DySPAN 2005)*, pp. 8-11

[24] B. Jashni, A. A. Tadaion, F. Ashtiani (2010). Dynamic link/frequency selection in multi-hop cognitive radio networks for delay sensitive applications, *IEEE 17th International Conference on Telecommunication (ICT 2010)*, pp. 128-132

5

Adaptation from Transmission Security (TRANSEC) to Cognitive Radio Communication

Chien-Hsing Liao and Tai-Kuo Woo
FooYin University/National Defence University
Taiwan (R.O.C.)

1. Introduction

Communication systems have to be made secure against unauthorized interception or against disruption or corruption in complicated electromagnetic environments. There are mainly three security categories used to delineate wireless communication systems generally, such as shown in Fig. 1, i.e., INFOSEC, COMSEC, and TRANSEC. We state about information security (INFOSEC) as that trying to against unauthorized access to or modification of information; we describe the communications security (COMSEC) as that keeping important communications secure. And we describe transmission security (TRANSEC) as that making it difficult for someone to intercept or interfere with communications without prior accurate waveforms, modulation schemes, and coding (Nicholson, 1987).

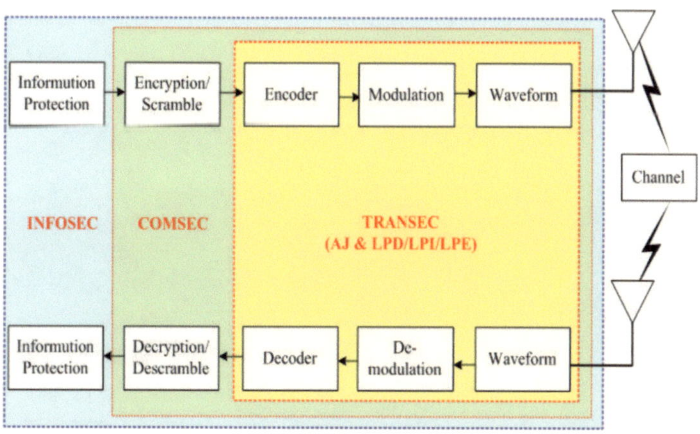

Fig. 1. Wireless communication security delineation

Therefore, in general, the basic strategies for acquiring and paralyzing the communication victim for the untended or intended users are to detect, intercept, exploit, and jam the communication signals. Relatively, the basic measures to counter these strategies for the victim are to design system with TRANSEC capabilities, i.e., low probability of detection, interception, and exploitation (LPD/I/E), and with receiving security capability, i.e., AJ or jam-resistant. LPD/I/E can be defined as measures with hidden signals which make it difficult for the unintended or intended receivers to detect between signal plus noise and noise alone, to distinguish between the signals, and to abstract feature and recover message, respectively. Nevertheless, the requirements for AJ communications are almost directly opposite to those for LPD/I/E communications. For example, whenever communication system increases power to maintain communication distance and counter a jammer, it will increase the threat of being detected or even intercepted by an unintended interceptor receiver accordingly. On the other hand, whenever communication system decreases power to counter the threat of an interceptor, it will also unavoidably decrease the AJ capability. The same is true if power is replaced with many other system related parameters, e.g., antenna size.

In addition, the receiving and transmission security achieved by a communication link depends very strongly on its location relative to an adversary's jamming transmitter and intercept receiver, which is categorized as geometry-dependent factors. In AJ applications, it is very important to have an accurate estimation of the processing gain required for reliable communications as a function of the link geometry. In cases of LPD/I/E, the detection probability is significantly affected by the interaction of the link geometry and the relative locations of the intercept receiver and the communication transmitter. In geometric point of view, the central concept of TRANSEC secure communications, is trying to force the related jamming, detection, interception, or exploitation measures, to have to work within our prescribed region(s), e.g., physically dead zones or lethal zones, when in comparison with conventional communications, as shown in Fig. 2. (Schoolcraft, 1991) The inner red region represents conceptually the maximum range necessary for acquiring the TRANSEC communications signatures by the adversary side when in comparison with conventional communications in the outer yellow region. Similarly for AJ concerns, jamming has to approach furthermore into TRANEC communications region as well.

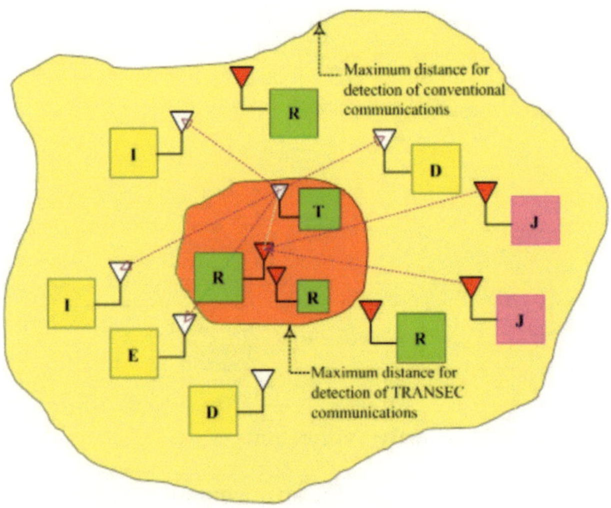

Fig. 2. TRANSEC communication scenario and concept

Under these circumstances as aforementioned, it is not straightforward to make wise and prudent evaluations and decisions for secure communications with concurrent AJ and LPD/I/E capabilities. Therefore, flexible and convenient metrics for achieving these are expected. Moreover, how to get an analysis model and metrics of evaluating effectively a special type of jammer with real-time (or near concurrent) detection (passive scanning) and transmission capability (called repeater or follow-on jammer) for a frequency hopping (FH) communication system is also expected. Finally, based on these, spectrum sensing technique like this with real-time detection and transmission capability, especially for a cognitive radio (CR) FH communication, is also expected to be adapted and available for communication resources sensing. This chapter is organized as follows.

In Section 2, a systematic approach for evaluating the interactions of the link geometry and TRANSEC system parameters (AJ and LPD are assumed) for secure communication is proposed. In many typical cases, communication designs have to deal simultaneously with adversary threats of both active jamming and passive detection to protect against jamming and detection. And by increasing communication power to counter jammer and enhancing anti-jamming capability must be weighted carefully against the increased threats of deteriorating low probability of detection capability and being detected by an unintended interceptor receiver. A qualitative and quantitative approach with sinc-type antenna patterns being included for evaluating both AJ and LPD concurrently is reached. And it is intuitive to see that by spreading signal spectrum, complicating signal waveforms, and lowering power control uncertainty, respectively, will enhance system security and performance accordingly.

In Section 3, a cognitive radio unit (CRU) model with uniform scanning (U-scanning) and sequential scanning (S-scanning) techniques and cognitive perception ratio (CPR) metric for cognitive communications adapted from TRANSEC is investigated. In this model real-time spectrum sensing characteristics are coordinated together with system parameters in temporal and frequency domains, e.g., scanning rate and framing processing time, for evaluating the performance of the cognitive radio (CR) communications under an elliptical or a hyperbolic operation scenario. CR technology has been proved to be a tempting solution to promote spectrum efficiency and relieve spectrum scarcity problems. Nevertheless, the cognitive capability cannot only be realized by monitoring on some frequency bands of interest but also more innovative techniques are required to capture the spectrum holes with temporal, frequency or spatial variations in sophisticated Frequency hopping spread spectrum (FHSS) radio environments, and avoid interference to the existing primary users. Nowadays, the FHSS systems have been widely used in civil and military communications, but somewhat their benefits would be potentially neutralized by a follow-on jamming (FOJ) with wideband scanning and responsive jamming capabilities covering the hopping period. The FOJ concept is actually implicitly analogous to a CR communication with spectrum and location awareness, listen-then-act, and adaptation characteristics. High CPR value means high spectrum awareness but low coexistence. Many intriguing numerical results are also illustrated to examine their interrelationships.

In Section 4, a systematic approach with their corresponding metrics for evaluating independent and concurrent AJ and LPD performance qualitatively and quantitatively is drawn in this section. Moreover, specific scanning schemes and a quantified CPR metric are available for evaluations of the coexistence of radio resources. Section 5 is literatures listing.

2. Secure communications system through concurrent TRANSEC evaluations

The central concept for a secure communications system is to protect against unintended or intended jammers and interceptors, force them to change system parameters or work outside of the prescribed acceptable regions, and maintain secure system performance simultaneously. In this section, a systematic approach for evaluating the interactions of the link geometry and TRANSEC system parameters for secure communication will be investigated.

2.1 Survey of related works

There are many inspiring methods and metrics which have already been explored and proposed by many forerunners for secure communication related performance concerns, which are addressed as follows. Turner described the reasons of LPD/LPI/LPE communications developments and anti-jamming verse LPD/LPI/LPE communications requirements, which offer capabilities not available with AJ communications. The ideal characteristics of a LPD/LPI/LPE communications waveform and methods for detecting LPD/LPI/LPE transmissions are listed to form the basis for discussing of their developments and capabilities (Turner, 1991). Glenn made a LPI analysis and showed the effect of scenario-dependent parameters and detectability-threshold factors in jamming and non-jamming environments, and concluded that the most significant improvement in LPI performance can be obtained by operating at Extremely High Frequency (EHF) and by maximizing the effective spread-spectrum processing gain and the communicator's antenna discrimination to the jamming signal, and by minimizing the number of symbols in the message (Glenn, 1983). Based on power gains and losses, Gutman and Prescott have given a LPI system quality factor (Q_{LPI}) to a grouping of quality factor terms consisting of the antennas (Q_{ANT}), type of modulation (Q_{MOD}), atmospheric propagation conditions (Q_{ATM}), and interference rejection capability (Q_{ADA}) in the presence of jammers and intercept receivers (Gutman & Prescott, 1989). Based on gain difference between the communication receiver and the radiometer, Dillards developed a detectability gain (DG) metric for defining "acceptable" LPD performance of a communication signal by a radiometer, which includes their path losses, antenna gains, etc., plus two "mismatch" losses incurred by the radiometers (Dillard & Dillard, 2001). Using classical radiometer analysis and communication theory, Weeks et. al. developed a methodology through detectability distance to quantify the LPD characteristics of some COTS wireless communication systems, i.e., GSM, IS-54, IS-95, and WCDMA, which is obtained by exploiting the step-like function behavior of the probability of detection curve for a given system. Tradeoffs between the observation time of the interceptor and the detectability distance with multiple users are also investigated (Weeks et al, 1998). Mills and Prescott presented a scenario-independent stand-off intercept model for situations in which the collocated network transmitters and the relatively distant interceptor are assumed. Under these assumptions, the detectability performance of the network using a frequency hopping multiple access scheme (FHMA) can be evaluated for the wideband and channelized radiometers and LPI quality factors can be used to compare the performance of them (Mills & Prescott, 1995). Furthermore, Mills and Prescott also established two multiple access LPI network detectability models, i.e., scenario-independent standoff network and scenario-dependent dispersed network models, and developed their corresponding LPI performance metrics to provide new insight into these issues (Mills & Prescott, 2000). Benia explored the effect of message length, processing gain, and coding gain on LPI performance and suggested a method for analyzing the effects of channel absorption loss models on the LPI quality factor (Binia, 2004). Wu investigated the impacts of the filter-bank interceptor performance and the emitter with FH waveform and specific antenna pattern on so-called circular equivalent vulnerable radius (CEVR) sensitivity within a statistical context like confidence interval (Wu, 2005).On the other hand, many forerunners have already investigated optimal interceptors for best detection probability concerns as well, which are addressed as follows. Schoolcraft defined six conventional and LPD waveforms to resist against seven detection techniques and provided a general approach to test the effectiveness of postulated threats against candidate waveforms and the relative LPD strengths of competing candidate waveforms (Schoolcraft, 1991). Wu studied the optimal interceptor for a FH-DPSK waveform, derived the detection algorithm based on the maximum likelihood principle, and proposed a novel performance evaluation approach (Wu, 2006). In spite of focusing on geometrical or power aspects of jamming only before, Burder analyzed and derived a mathematical intercept model for computation of the jamming probability when a follower jammer with a wideband-scanning receiver jams a single FH system (Burda, 2004). Gross and Chen developed and predicted the relative detection range for two types of transmitted waveforms, i.e., a benchmark rectangular pulse and a Welti binary coded waveform, and some classic passive receivers, e.g., square-law, delay and multiply, wideband, and channelized receivers possessing typical bandwidth, noise-floor, and loss parameters (Gross & Chen, 2005). In spite of the active jamming measures taken, FOJ is implicitly analogous to a cognitive radio communication with spectrum and location awareness, listen-then-act, and adaptation characteristics. For transmission security concerns, concurrent anti-jamming and low probability detection were investigated to have a secure communication (Liao et al., 2007).

2.2 System analysis scenario

The operation scenario with both AJ and LPD for a Ka-band GEO satellite communication system will be taken as system analysis model in this subsection. The relative geometry locations of the victim satellite communication system and the adversary multiple jammers and interceptors are shown in Fig. 3, where the victim communication terminal is put at the origin, and the latter two are collocated on the same fixed or varying positions, i.e., (x_1, y_1, 0), (x_2, y_2, 0), (x, y, 0), and (x_N, y_N, 0) etc. The paired jammer and interceptor on the same position (x, y, 0) are varying in an approaching or receding way for evaluating AJ and LPD performance. R_c is the range between our communication system themselves (e.g., R_c is assumed to be 36000kM for a geosynchronous elliptic orbit (GEO) satellite). R_j/R_i is the jammer/interceptor range from their collocated earth position to the victim communicator on the position of (0, 0, R). As shown in this figure, one very important factor for the effectiveness of jamming or interception is the relative angle φ off the main beam pattern of the victim communicator (on the position of (0, 0, R)) in the direction of earth jammer/interceptor. Generally, our friendly communication system (communicators at the origin and on the position (0, 0, R)) will direct main beam patterns at each other to get maximum gain patterns, and the intentional jammers or unintended interceptors will only be able to cover from the sidelobe direction, especially when the operating frequency is higher. Fig. 3 shows a basic scenario with an intercepted and jammed satellite and multiple jammers and interceptors on the ground. Whenever the jamming and interception range is very far away from the victims, e.g., a GEO satellite, the interceptor and jammer will also get the advantage of the main beam gain of the victim communicator on the position (0, 0, R) relatively easier, if no antenna shaping or nulling pattern designs are taken.

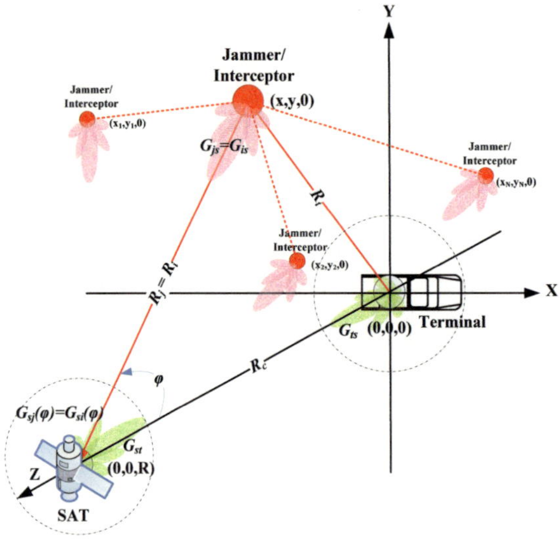

Fig. 3. The operation scenario with both AJ and LPD for a Ka-band GEO satellite communication system

Jammer/interceptor will get the advantages of the victim antenna pattern (mainlobe) to enhance their jamming or interception effects if the relative angles of the jammer/interceptor and the victim communicator on the position (0, 0, R) are tilted with the line of sight of the victim communicators themselves. The well-known parabolic antenna power gain $G_o(D)$ and pattern $G(\varphi)$ are given as follows (Jain, 1990)

$$G(\varphi)=\eta(\pi D/\lambda)^2\left(\frac{2J_1(\pi D\sin(\varphi)/\lambda)}{\pi D\sin(\varphi)/\lambda}\right)^2, \tag{1}$$

where D is antenna diameter, λ is wavelength, η is antenna efficiency, and J_1 is first order Bessel function. If the satellite communicator on the position of (0, 0, R) is assumed to be the victim with a 3m sinc-type antenna, a narrower 30GHz and a wider 20GHz antenna patterns will be for uplink and downlink communication for the terminals on the ground, respectively. It is therefore good for anti-jamming design (30GHz) but not for LPD (20GHz) due to its wider transmission pattern.

2.3 Evaluation method

Not only are the power control schemes of a communication system very crucial to provide adaptive adjustments of transmission power and signal-to-noise ratios required, but are also the communication signal very dangerous to be more easily detected by the unintended or intended interceptors, especially, whenever the power sources are transmitted unscrupulously. In general, the receiving and transmission security achieved by a communication link depends very strongly on its system related parameters and location relative to an adversary's jamming transmitter and intercept receiver, which can be categorized as system-dependent and geometry-dependent factors, respectively, for independent or concurrent AJ and LPD security concerns. The jamming to signal ratio (J/S) and the intercept signal to noise ratio ($(S/N)_i$) can be categorized, respectively, as system-dependent factors, i.e., f(•) and g(•) and geometry-dependent factors, i.e., R_{js}(•) and R_{si}(•), respectively. In coordination with the system operation scenario as shown in Fig. 3, the signal bit energy to jamming density (E_b/J_o), where J_o is very large compared to noise power density N_o, can be categorized as system-dependent factor h(•) and geometry-dependent factors R_{ej}(•), respectively. Furthermore, E_b/J_o can be proportionally or inversely related to these four system related parameters, i.e., processing gain (G_P), effective jamming power for the adversary jammer (ΔM_j), effective intercepting sensitivity for the adversary interceptor (ΔM_i), and effective power control for the victim communication system (ΔM_t). For example, whenever any of the latter three parameters is increased, E_b/J_o is lowered, i.e., error probability is increased. In contrast, whenever any of these three parameters is decreased, error probability is decreased.

2.4 Concurrent AJ and LPD communications (E_b/J_o)

In this subsection, in order to consider the AJ and LPD applications concurrently, we assume that the jamming power spectral density J_o is very large compared to noise power density N_o. Binary frequency shift keying (BFSK) is a kind of digital modulation method with two frequencies representing 0 and 1. Frequency hopping multiple access (FHMA) is a kind of spread spectrum schemes with many orthogonal or pseudo-orthogonal frequency patterns for multiple users application. The error probability for BFSK and FHMA combination, P_{fh}, is given by (Sklar, 2001)

$$P_{fh} \triangleq \frac{1}{2} \cdot e^{-\frac{1}{2}\left(\frac{E_b}{J_o}\right)}, \tag{2}$$

where E_b (=S/R_b) and J_o are, respectively, the bit energy and jamming power density given as follows for the nth paired jammer/interceptor

$$J_{o,n} = P_{j,n} G_{js,n} G_{sj,n}(\varphi) \left(\frac{\lambda}{4\pi R_{j,n}} \right)^2 \frac{1}{W_{ss}} \qquad J_{o,n} \gg N_o, \tag{3}$$

$$E_{b,n} = \frac{S_{t,n}}{R_b} = \left(\frac{S}{N} \right)_{io} \frac{G_{ts} G_{st}}{G_{ti,n}(\varphi) G_{it,n}} \frac{L_{i,n}}{L_t} \frac{N_{oi,n} W_{ss}}{R_b} \left(\frac{R_{i,n}}{R_c} \right)^2 \tag{4}$$

By substituting equation (3) and (4) into equation (2) and taking an inverse natural logarithm of it, the required E_b/J_o for the nth paired jammer/interceptor is given as follows

$$\left(\frac{E_b}{J_o} \right)_n = \left(\frac{4\pi}{\lambda} \right)^2 \left(\frac{S}{N} \right)_{io} \left(\frac{G_{ts} G_{st} L_{i,n} N_{oi,n}}{G_{is,n} G_{js,n} G_{si,n}(\varphi) G_{sj,n}(\varphi)} \right) \frac{W_{ss}^2}{P_{j.n} R_b} \left(\frac{R_{i,n} R_{j,n}}{R_c} \right)^2, \tag{5}$$

$$\left(\frac{E_b}{J_o}\right)^{-1} = \sum_{n=1}^{N}\left(\frac{E_b}{J_o}\right)_n^{-1} \triangleq h(\bullet)R_{ej}(\bullet), \tag{6}$$

where $(S/N)_{io}$ means the optimum intercepting signal-to-noise ratio of an interceptor receiver; N_{oi} represents the output thermal noise power density of interceptor receiver, which is equal to $kT_0\ N_{fi,n}$, where k, T_0 and $N_{fi,n}$ are Boltzmann's constant (1.38×10^{-23} W/Hz/°K), absolute temperature (290°K assumed), and interceptor receiver noise figures, respectively. Equation (5) can also be categorized as system-dependent parameter h(•) and geometry-dependent parameters $R_{ej}(\bullet)$ as shown in equation (6). The E_b/J_o can be further manipulated and given as follows

$$\frac{E_b}{J_o} = \frac{(S/N)_{io}(S/N)_{to}G_P}{(J/S)(S/N)_i(S/N)_t}, \tag{7}$$

where G_p is the spreading spectrum processing gain given by W_{ss}/W_{bb} and $(S/N)_t$ is be given as follows

$$\left(\frac{S}{N}\right)_t = \frac{P_t G_{ts} G_{st}\left(\frac{\lambda}{4\pi|R_c|}\right)^2}{kT_o N_{ft} W_{bb} L_t} \tag{8}$$

From equation (7), J/S is inversely related to $(S/N)_i$ with the other parameters fixed, which can be further manipulated and simplified as given by

$$\frac{E_b}{J_o} = \frac{G_P}{\Delta M_j\ \Delta M_i\ \Delta M_t}, \tag{9}$$

where ΔM_j, ΔM_i, and ΔM_t are defined as J/S, $(S/N)_i/(S/N)_{io}$, and $(S/N)_t/(S/N)_{to}$, respectively. They mean the ratios of effective jamming power for the adversary jammer, effective intercepting sensitivity for the adversary interceptor, and effective power control for the victim communicator. E_b/J_o is inversely related to these three parameters if G_p is fixed. For example, whenever any of these three parameters is increased, E_b/J_o is lowered, i.e., error probability is increased. On the contrary, whenever any of these three parameters is decreased, error probability is decreased. From equation (9), it is intuitive for the victim communicator side to enhance system performance (lowering error probability) by spreading signal spectrum (G_p is increased), increasing signal power (ΔM_j is decreased), complicating signal waveforms (ΔM_i is decreased), and (or) lowering power control uncertainty (ΔM_t is decreased), respectively. For the adversary side, in order to deteriorate the performance of this secure victim communication system, ΔM_j, ΔM_i, and ΔM_t should be increased to lower E_b/J_o accordingly. In fact, the proposed concurrent AJ and LPD research can contribute to the CR communications for practical implementation by replacing all the collocated jammers/interceptors as shown in Fig. 3 with "cooperative" communicators, in which some geometry- and system-dependent factors can be sensed and aware of for spectrum resources adjustments or accesses, e.g., ΔM_j, ΔM_i, ΔM_t, relative positioning locations, and their corresponding parameters like power, bandwidth, spectrum hole, and etc.

2.5 E_b/J_o numerical analysis

In this section, one typical Ka-band (f_{up}/f_{dn}=30/20GHz) GEO satellite communication (SATCOM) example (R_c is assumed to be 36000kM) is illustrated through the proposed design approach for a secure communication system with concurrent requirements of both AJ and LPD capabilities. As shown in Fig. 3, the 7-m size jammer and interceptor antennas with maximum sinc type antenna pattern gains of G_{js} and G_{is}, respectively, are assumed in simulations. The antenna sizes for the communicators at the origin and on the position (0, 0, R) are both 3-m, and they are pointed each other with maximum gains. The sidelobe patterns leaked from the victim communicator on the position (0, 0, R) (i.e., GEO satellite) to the collocated interceptor and jammer on the ground are tilted φ angle dependent on relative positions among them, i.e., $G_{si}(\varphi)$ and $G_{sj}(\varphi)$. Based on equation (8) and (9) for concurrent AJ and LPD considerations, their corresponding figures of E_b/J_o verse R_t range for single or multiple collocated jammers/interceptors are shown in Fig. 4.

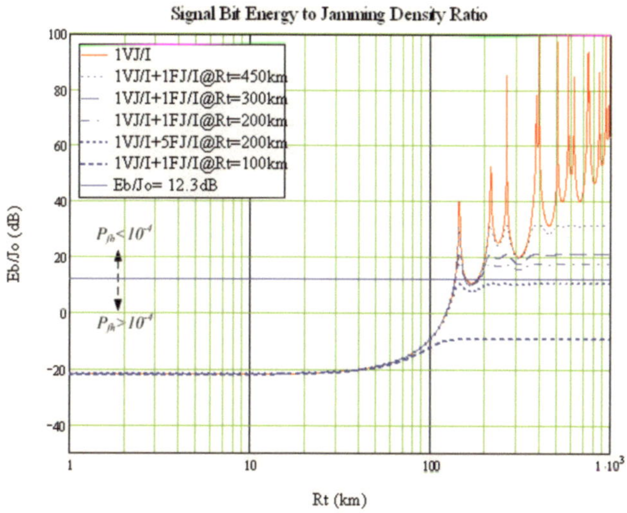

Fig. 4. Six different E_b/J_o ratios comparison with one varying collocated jammer/interceptor (VJI) and multiple collocated fixed jammer/interceptor (FJI) combinations

From Fig. 4, six different E_b/J_o ratios comparison with one varying collocated jammer/interceptor (VJI) and multiple collocated fixed jammer/interceptor (FJI) combinations are shown: 1 VJI(red solid—), 1 VJI + 1 FJI @R_t=450km(blue dot·····), 1 VJI + 1 FJI @R_t =300km(blue dash----), 1 VJI + 1 FJI @R_t=200km(blue dash dot-·-·-), 1 VJI + 5 FJI @R_t=200km(thick blue dot·····), and 1 VJI + 1 FJI @R_t=100km(thick blue dash----). The concurrent AJ and LPD performance under these single or multiple jammers/interceptors operation scenario can be examined if a minimum 10^{-4} FH error probability (P_{fh}) is asked to maintain (E_b/J_o=12.3dB). It is clear that whenever one single paired jammer/interceptor is beyond the range (R_t≈139kM), the communication performance criterion will be met ($P_{fh}<10^{-4}$) and not affected by this threat. Nevertheless, whenever under this operation scenario with one varying paired jammer/interceptor plus five jammers/interceptors all at R_t=200km, the specified communication criterion can not be met any more. For multiple fixed paired jammers/interceptors with one varying jammer/interceptor concerns, we find that there exist even more "smoothed" effects for E_b/J_o curves when in comparison with J/S and $(S/N)_i$ curves as aforementioned, which maybe are due to randomized sidelobe patterns effects.

Fig. 5 shows the respective intuitive E_b/J_o contour plot with one varying jammer/ interceptor and one collocated fixed jammer/interceptor (1FJI) at R_t=200km. The prescribed E_b/J_o=12.3dB circle (radius is about 139km) within which $P_{fh}>10^{-4}$ is also shown for concurrent AJ and LPD performance evaluations criterion. Therefore, a specific strategy could be taken to expel the approached varying jammer/interceptor beyond this zone. A much more smoothed E_b/J_o contour outside the prescribed criterion circle is observed, which means a less secure communication performance even only one more fixed jammer/interceptor at distant range is considered. According to equation (9), four main system-dependent factors can be as metrics for tradeoffs.

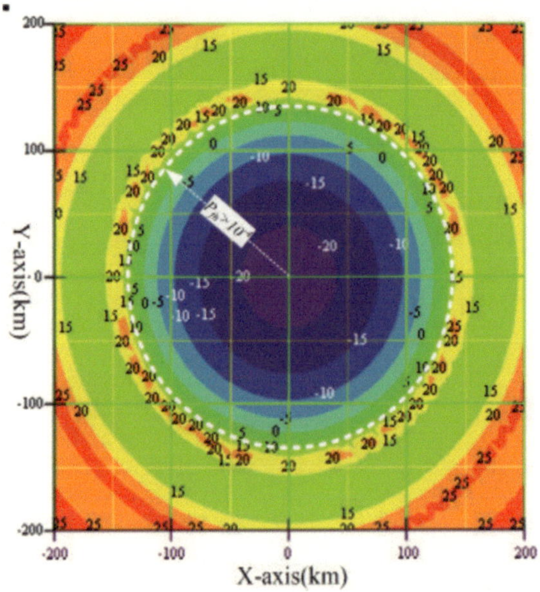

Fig. 5. E_b/J_o contour plot with one collocated varying jammer/interceptor (1VJI) and one collocated fixed jammer/interceptor (1FJI)

3. Spectrum sensing capability through specific scanning schemes

In previous section, we have proposed the approach of TRANSEC design by considering both AJ and LPE capabilities concurrently for multiple jammers and interceptors. In this section, we will further investigate a kind of special jamming with both real-time scanning and transmission characteristics, i.e., concurrent detection and jamming capabilities, which is designed inherently to counter a FH communication system. This should be a good bench target to evaluate the performance of the FH spread spectrum system with both transmission and reception characteristics being considered simultaneously. Moreover, effective jamming probability metrics for specified scanning schemes taken by FOJ, are investigated, which will be good figures of merit for evaluating FH and CR communication system performance.

3.1 Survey of related works

For the past years, traditional spectrum management approaches have been challenged by their actually inefficient use or low utilization of spectrums even with multiple allocations over many of the frequency bands (NTIA). Thus, within the current regulatory frameworks of communication, spectrum is a scarce resource ,Spectrum policy task force report, 2002). Cognitive radio is the latest emphasized technology that enables the spectrums to be used in a dynamic manner to relieve these problems. The term "cognitive radio (CR)" was first introduced in 1999 by Mitola and Maguire and is recognized as an enhancement of software defined radio (SDR), which could enhance the flexibility of personal wireless services through a new language called the *radio knowledge representation language* (RKRL), and the cognition cycle to parse these stimuli from outside world and to extract the available contextual cues necessary for the performance of its assigned tasks (Mitola III & Maguire Jr., 1999; Mitola III, 2000). Haykin therefore defines the cognitive radio as an intelligent wireless communication system that is aware of its surrounding environment, and uses the methodology of understanding-by-building to learn from the environment and adapt its internal states to statistical variations in the incoming *RF stimuli* by making corresponding changes in certain operating parameters in real-time (Haykin, 2005). In addition, some engineering views and advances for helping the implementation of cognitive radio properties into practical communications are described (Jondral & Karlsruhe, 2007; Mody et al., 2007). With these groundbreaking investigations and developments, international standardization organizations and industry alliances have already established standards and protocols for cognitive radio as well (Cordeiro et al., 2005; Ning et al., 2006; Cordeiro et al., 2006). The frequency hopping spread spectrum (FHSS) systems are widely used in civil and military communications, but somewhat the benefits of FHSS systems could be potentially neutralized by a follow-on jamming (FOJ) with an effective jamming ratio covering the hopping period (Torrieri, 1986; Felstead, 1998; Burda, 2004).

In spite of the active jamming measures taken, FOJ is implicitly analogous to a cognitive radio communication with spectrum and location awareness, listen-then-act, and adaptation characteristics. Therefore, the cognitive process cannot be simply realized by monitoring the power or signal-to-noise ratio in some frequency bands of interest in a FH radio environment. For transmission security concerns, concurrent anti-jamming and low probability detection were investigated to have a secure communication (Liao et al., 2007). Furthermore, real-time spectrum sweeping characteristics are coordinated together with system parameters in temporal and frequency domains, e.g., scanning rate and framing processing time, for evaluating the performance of CR communications under an elliptical or a hyperbolic operation scenario, which can be applied for radio spectrum sensing and location awareness in cognitive radio communications. The proposed schemes and metrics can pave one practical way for system evaluations of cognitive radio communications (Liao et al., 2009). Although the performance evaluation of cognitive radio (CR) networks is an important problem, it has received limited attention from the CR community. It is imperative for cognitive radio network designers to have a firm understanding of the interrelationships among goals, performance metrics, utility functions, link/network performance, and operating environments. Various performance metrics at the node, network, and application levels are reviewed. A radio environment map-based scenario-driven testing (REM-SDT) for thorough performance evaluation of cognitive radios and an

IEEE 802.22 WRAN cognitive engine testbed are also presented to provide further insights into this important problem area, respectively (Zhao et al., 2009). A coexistence window inside which the primary user and secondary user share the radio channel in time division manner is proposed. Connectivity probability and link utility efficiency are defined to measure the performance of secondary user. Considering the practical noise channel, how the metrics change is studied and the data rate of the secondary user in this case is obtained (Liu et al., 2010). An optimal power allocation scheme for a physical layer network coding relay based secondary user (SU) communication in cognitive radio networks is proposed. SUs are located on two different primary user (PU) coverage areas and an energy and spectrally efficient SU communication scheme is introduced (Jayasinghe et al., 2011). Finally, a latest systematic overview on CR networking and communications by looking at the key functions of the physical (PHY), medium access control (MAC), and network layers involved in a CR design and how these layers are crossly related are proposed, which can help researchers and practitioners have a clear cross-layer view on designing CRNs (Liang et al., 2011).

3.2 Model for FH jamming probability

Fig. 7 shows the basic FOJ function block diagram with many allocated process time to acquire incoming "victim" signals and implement jamming, where jT_z time is the total analytical time needed to acquire the instant hoping frequency, τ_r is the total activation time needed to synthesize and amplify a repeater signal tone or noise to jam the "victim" signal, which may compose of the process times of frequency synthesizer, power amplifier, and filter banks.

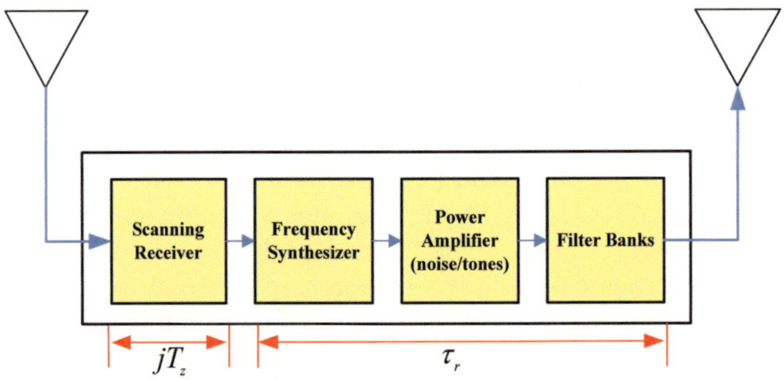

Fig. 6. Basic FOJ function block diagram

In general, the processed times for these three parameters are in the order of us, ns, and ns, respectively. Furthermore, the propagation delay or difference time (τ_d) dependent on relative positions should be included for effective jamming probability analysis. For example, if the range difference (ΔR) is 30km, the τ_d propagation difference time will be around 100 μs, far longer than the response time τ_r. Therefore, this parameter can be assumed to be zero while compared to other larger parameters under this circumstance.

Fig. 8 shows the effective jamming dwell time breakdown for FOJ, where T_r represents the jamming total delay time of process delay and propagation time ($=\tau_r+\tau_d$), T_l represents the latency time ($=jT_z + T_r$), and T_j represents the effective jamming dwell time ($=T_h-T_l$).

$$T_l = jT_z + (\tau_r + \Delta\tau_d) = jT_z + T_r , \tag{10}$$

$$T_J = T_h - (jT_z + T_r) = T_t - jT_z \tag{11}$$

T_j must be smaller than T_h under any circumstance. The effective jamming probability (h) is defined to be the ratio of T_j and T_h.

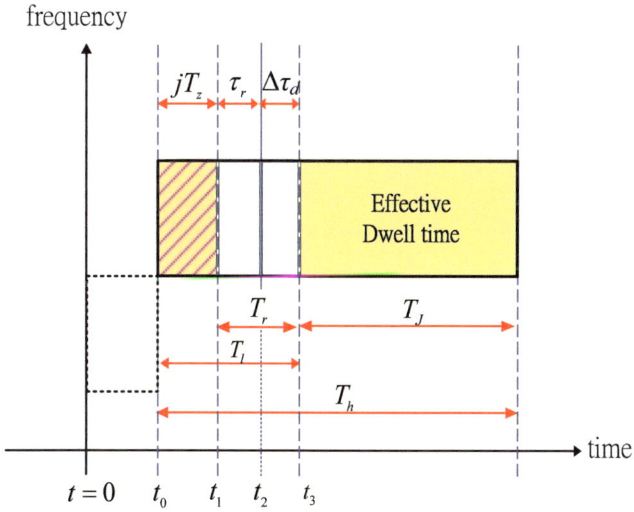

Fig. 7. Effective dwell time (T_J) and latency time breakdown for CRU operation

Scanning windows available during hopping dwell interval is defined to be m, which is represented as

$$m = \left\lfloor \frac{T_h - T_r}{T_z} \right\rfloor = \left\lfloor \frac{T_t}{T_z} \right\rfloor , \tag{12}$$

where T_z represents the analysis framing time per scanning window W_s of the jammer and $\lfloor x \rfloor$ symbol means the maximum integer equal to or smaller than x. It follows that the follower jammer is able to analyze at most m scan windows during the single dwell interval, T_h. Scanning window number available in the FH system bandwidth is defined to be n and represented as

$$n = \left\lceil \frac{W}{W_s} \right\rceil , \tag{13}$$

where W represents the hopping bandwidth of a FH system, W_s represents scanning window of the jammer, and $\lceil x \rceil$ symbol means the minimum integer equal to or larger than x. Let k be the number of scan windows which the FOJ analyzes in the dwell interval. It is evident that

$$k = \min(m, n) , \tag{14}$$

which means the smaller one of m or n is selected as the analyzed number of scanned windows. Let us suppose that one FH system operates in the bandwidth W only and that the follower jammer knows the parameters of the FH system and knows the moments of channel changes as well. Therefore, exactly at these moments (t=0), the jammer will initiate searching of the actual channel. When FH terminal transmits in j^{th} scan window, then this transmission is found at moment $t_0 = jT_z$. Let t_1 be the moment when the scanning receiver finds the actual transmission channel of the FH system. Let t_2 be the moment when the follower jammer initiates jamming of the found channel. Let t_3 be the moment when the initiated signal of the FOJ reaches the receiver site of the found channel, i.e., the FH receiver is jammed at the moment t_3.

3.3 Scanning schemes

In the following sections, two schemes named uniform scanning and sequential scanning will be explored and taken as scanning measures to scan and trace the incoming hopping signals fast enough to implant effective noise or tone jamming thereafter. In addition, the case of delay response ($T_r \neq 0$) will be examined as well for these two scanning schemes.

3.3.1 Uniform scanning (U-scanning) scheme

A uniform scanning (U-scanning) technique will be explored and taken as the scanning measures to scan and trace the incoming hopping signals fast enough to implant transmission signals thereafter. If the CRU analyzes all scan windows randomly with uniform probability $p_u(T_J)=1/n$, and $p_u(T_J)=(n-k)/n$ is the probability that the FH system operates in the scanning window which is not analyzed. Therefore, the probability distribution of the jammed period of the dwell interval can be given by

$$p_u(T_J) = \begin{cases} \dfrac{(n-k)}{n}, & j > k\left(T_J = 0\right) \\ \dfrac{1}{n} & , j = 1,2,...k \end{cases} \tag{15}$$

It is assumed that T_r is assumed not to be zero, i.e., τ_r is zero, but τ_d is not zero and $T_r=\tau_d=l\times T_h$, where l is the propagation time ratio between $T_r = \tau_d$ and T_h. The average jammed period of the dwell interval is therefore derived and given by

$$\overline{T}_{Ju} = \sum_{j=1}^{n} T_J \cdot p_u\left(T_j\right) = \frac{k}{n}\left((1-l)T_{hu} - T_z\frac{k+1}{2}\right) \tag{16}$$

From the above derived equation, the criterion of hopping rate (R_{hu}) and analysis framing time product (T_z) for effective dwell can be available and given by

$$R_{hu} \cdot T_z \leq (1-l) \cdot \frac{2}{(k+1)}, \tag{17}$$

which is the basic condition whenever $T_r \neq 0$ for effective CRU. The effective dwell ratio and scanning rate (R_{su}) for uniform scanning technique can be expressed and given by equation (18) and (19), respectively.

$$h_u = \frac{\overline{T}_J}{T_h} = \frac{k}{n} \cdot \left((1 - l) - \frac{T_z}{T_h} \frac{k+1}{2} \right), \tag{18}$$

$$R_{su} = \frac{W_s}{T_z} = W_s \cdot R_{hu} \cdot \left(\frac{k(k+1)}{2 \cdot ((1 - l) \cdot k - nh_u)} \right) \tag{19}$$

3.3.2 Sequential scanning (S-scanning) scheme

A sequential scanning scheme will be taken as the scanning measure to scan the incoming frequency hopping signals fast enough to implant CRU transmit signal if it is allowable. Based on the basic definitions as aforementioned, if CRU analyzes all scanning windows randomly with sequential perception $p_s(T_J)=1/(n+1-j)$, then $p_s(T_J)=(n-k)/(n+1-j)$ will be the perception not analyzed in the scanning window. Therefore, the perception distribution of the effective dwell time can be given by

$$p_s(T_J) = \begin{cases} \dfrac{n-k}{n+1-j}, & j > k(T_J = 0) \\ \dfrac{1}{n+1-j}, & j = 1,2,\ldots,k \end{cases} \tag{20}$$

It is assumed that T_r is assumed not zero and $T_r=\tau_r+\Delta\tau_d=l\times T_h$, where l is the propagation time ratio between T_r and T_h. The average effective dwell time can therefore be derived and given by

$$\overline{T}_{Js} = \sum_{j=1}^{k} T_J \cdot p_s(T_j) = \sum_{j=1}^{k} \left(\frac{(1-l) \cdot T_{hs} - jT_z}{n+1-j} \right) \tag{21}$$

From (21), the criterion of hopping rate ($R_{hs}=1/T_{hs}$) and framing processing time product (T_z) for effective dwell time can be available and given by

$$R_{hs} \cdot T_z \le \frac{\sum_{j=1}^{k} \left(\frac{1-l}{n+1-j} \right)}{\sum_{j=1}^{k} \left(\frac{j}{n+1-j} \right)}, \tag{22}$$

which is the basic criterion whenever $T_r \neq 0$ for effective coverage of the hopping period. Therefore, the effective dwell time ratio and the scanning rate by sequential scanning scheme can be manipulated further and given by equation (23) and (24), respectively.

$$h_s = \frac{\overline{T}_J}{T_h} = \sum_{j=1}^{k} \left(\frac{1-l}{n+1-j} \right) - \frac{T_z}{T_{hs}} \cdot \sum_{j=1}^{k} \left(\frac{j}{n+1-j} \right), \tag{23}$$

$$R_{sh} = \frac{W_s}{T_z} = W_s \cdot R_{hs} \cdot \left(\frac{\sum_{j=1}^{k} \left(\frac{j}{n+1-j} \right)}{\sum_{j=1}^{k} \left(\frac{1-l}{n+1-j} \right) - h_s} \right) \tag{24}$$

3.4 Geometric model for FH jamming

In this section, two analytical geometric models, i.e., elliptic and hyperbolic, will be examined furthermore dependent on their relative positions among the FOJ, FH transmitter, and FH receiver.

3.4.1 Elliptical FH jamming model

Fig. 9 shows an elliptic FH jamming model. If the relative positions among the FOJ, FH transmitter, and FH receiver are shown in Fig. 9 with fixed range R_{tr}=a between the FH transmitter and receiver and varying FOJ position, then the following expressions will be available by using the fact that latency time (T_l) should be smaller than the hopping period (T_h) for an effective jamming.

$$T_l = jT_z + \tau_r + \tau_d = jT_z + \tau_r + \frac{\left(R_{tj} + R_{jr} - a\right)}{c} \leq T_h \,, \tag{25}$$

where τ_r can be assumed to be zero for instant response for the jammer, R_{jr} is the range between transmitter and FOJ, and R_{jr} is the range between FOJ and receiver.

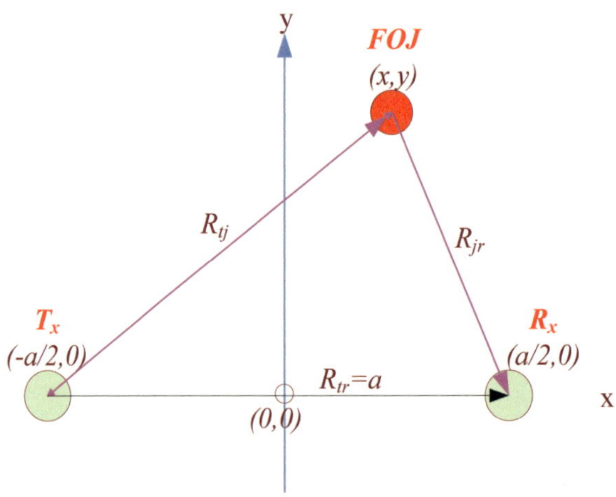

Fig. 8. Elliptic FH jamming model

After a simple manipulation, a standard ellipse equation will be given by

$$\frac{x^2}{\left(D+a\right)^2} + \frac{y^2}{D \times \left(D+2a\right)} \, {}^{3} \frac{1}{4} \,, \tag{26}$$

where D is assumed to be given by

$$\left(R_{tj} + R_{jr} - a\right) \leq \left(T_h - jT_z - \tau_r\right) \cdot c = D \tag{27}$$

Fig. 10 shows the typical hopping rate (R_h) contours for an elliptic FH jamming model with varying FOJ locations and assumed scanning time around 1ms (jT_z=10×100μs) and fixed R_{tr}=a =100km. Whenever a specified fixed hopping rate is required (e.g. R_h=500Hz), an even higher hopping rate is necessary for the FH communication system if FOJ is penetrated through this boundary and inside the specified ellipse. On the contrary, if FOJ is located outside the ellipse boundary, then the specified hopping rate is fast enough to counter the FOJ jamming for the FH communication system.

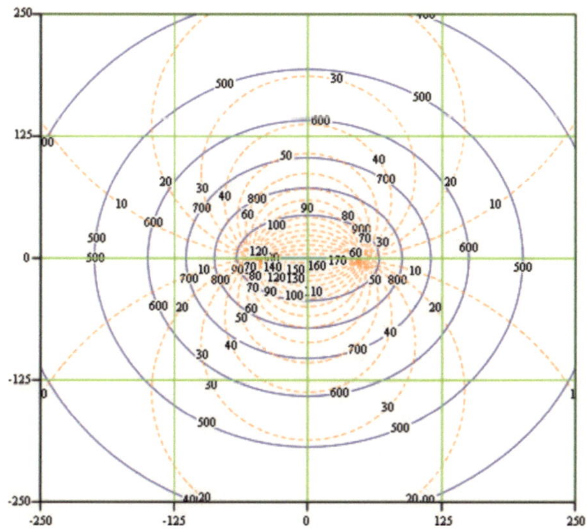

Fig. 9. Typical elliptic contour of FH jamming model

3.4.2 Hyperbolic FH jamming model

If the relative positions among the FOJ, FH transmitter, and FH receiver are shown in Fig. 11 with fixed range R_{tj}=a between the FH transmitter and FOJ and varying FH receiver position, then the following expressions will be available by using the fact that latency time (T_l) should be smaller than the hopping period (T_h) for an effective jamming.

$$T_l = jT_z + \tau_r + \tau_d = jT_z + \tau_r + \frac{\left(a + R_{jr} - R_{tr}\right)}{c} \leq T_h, \tag{28}$$

where τ_r can be assumed to be zero for instant response for the jammer, R_{jr} is the range between FOJ and receiver, and R_{tr} is the range between transmitter and receiver. After a simple manipulation, a standard ellipse equation will be given by

$$\frac{x^2}{\left(D - a\right)^2} - \frac{y^2}{D \cdot \left(2a - D\right)} \geq \frac{1}{4}, \tag{29}$$

where D is assumed to be given by

$$\left(a+R_{jr}-R_{tr}\right) \leq \left(T_h-jT_z-\tau_r\right) \cdot c=D, \tag{30}$$

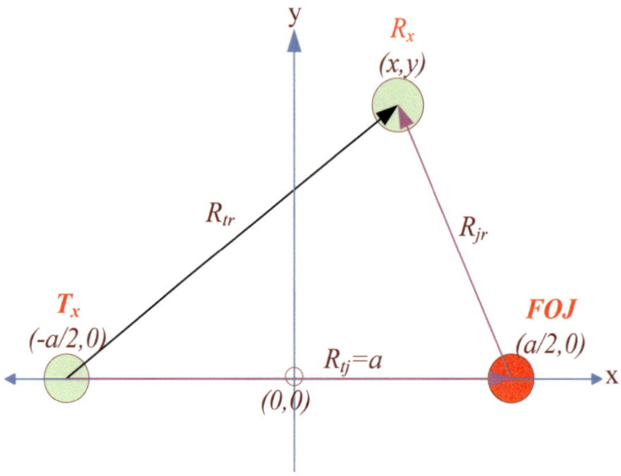

Fig. 10. Hyperbolic FH jamming model

Fig. 12 shows the typical contours for a hyperbolic FH jamming model with varying FH receiver locations and assumed scanning time around 1ms (jT_z=10×100μs) and fixed R_{tj}=a=100km. Whenever a specified fixed hopping rate is required (e.g. R_h=650Hz) under this circumstance, an even higher hopping rate is necessary for the FH communication system if the varying FH receiver is penetrated through this boundary to be closer to the fixed FOJ position on the right side. On the contrary, if the varying FH receiver is located on the left side of the hyperbolic boundary, then the specified hopping rate is fast enough to counter the FOJ jamming for the FH communication system.

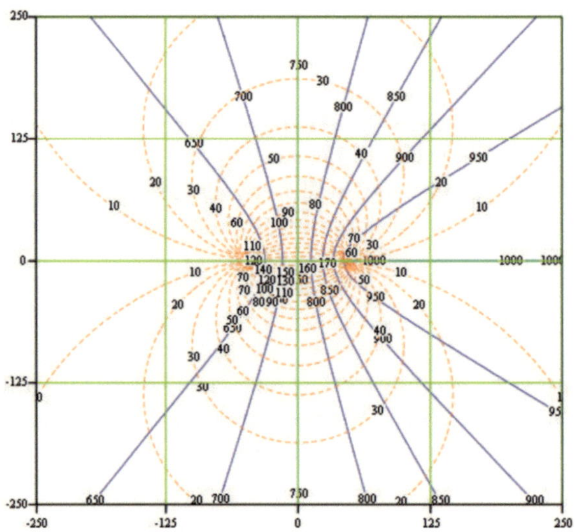

Fig. 11. Typical hyperbolic contour of FH jamming model

Fig. 13 shows the comparison results of effective jamming probability (h) vs. hopping rate (R_h) for both uniform (U-) scanning and sequential (S-) scanning schemes. Under the same framing time (T_z) conditions it is observed obviously that S-scanning scheme is better than U-scanning scheme for fixed hopping rate, e.g., the effective jamming probability value will be around 0.8 and 0.5 if R_h=500Hz and T_z= 100us for S- and U-scanning scheme, respectively. In another point of view, S-scanning scheme will have better hopping rate sensing capability (650Hz) than U-scanning scheme (500Hz) if effective jamming probability is fixed at 0.5.

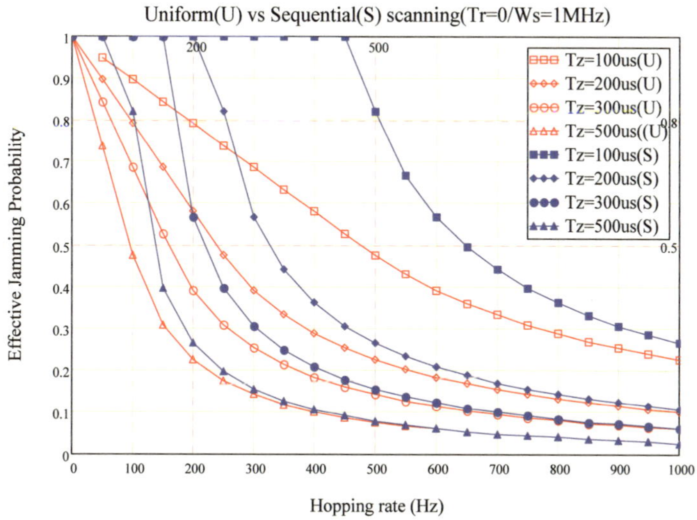

Fig. 12. Effective jamming probability vs. hopping rate with specified T_z values

3.5 Cognitive radio unit (CRU) and cognitive perception ratio (CPR)

Fig. 7 shows the basic FOJ function block diagram with many allocated process time to acquire incoming "victim" signals and implement jamming. This function block diagram is basically analogous to a cognitive radio unit (CRU) for sensing spectrum signals while applied in a cognitive communications adapted from TRANSEC. In addition, CRU models with U-scanning and S-scanning schemes and cognitive perception ratio (CPR) metric for quantified cognitive communications could be available as well. In this model real-time spectrum sensing characteristics can be coordinated together with system parameters in temporal and frequency domains, e.g., scanning rate and framing processing time, for evaluating the performance of CR communications under an elliptical or a hyperbolic operation scenario. Fig. 14 and Fig. 15 show the hyperbolic and elliptic CPR contours, respectively, with both U- and S-scanning being put together for comparisons, and with T_z=100us and R_h= 500Hz.

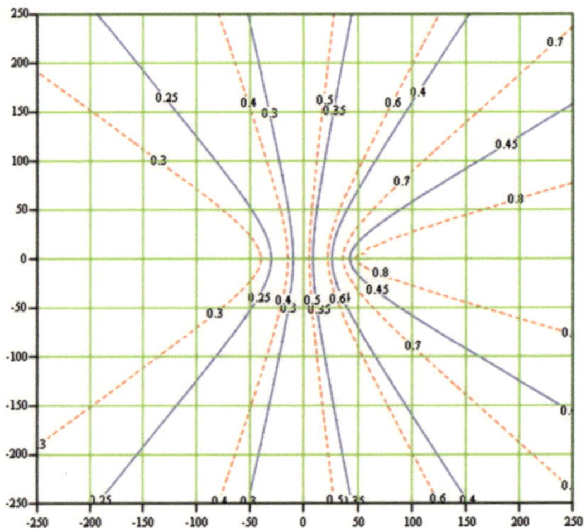

Fig. 13. Hyperbolic CPR contours with U- and S-scanning & R_h= 500Hz

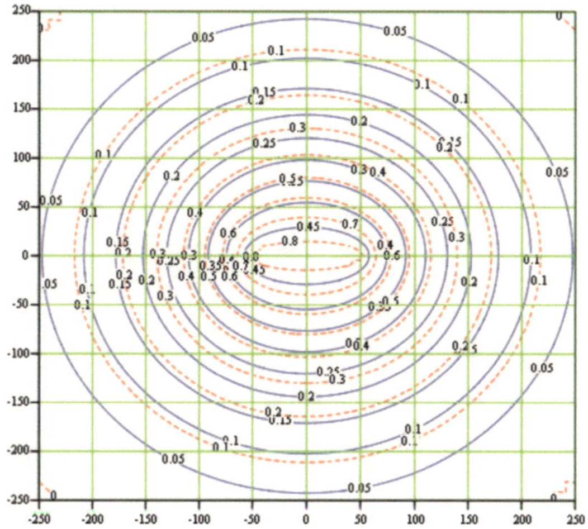

Fig. 14. Elliptic CPR contours with U & S scanning & R_h= 500Hz

It is observed from Fig. 14 that the CPR-contour values of 0.25 to 0.45 (blue solid) and values of 0.3 to 0.8 (red dashed) for U-scanning and S-scanning, respectively, are shown from the left hyperbolic trajectories to the right trajectories. And it is observed from Fig. 15 that the CPR-contour values of 0.05 to 0.45 (blue solid) and values of 0.1 to 0.8 (red dashed) for U-scanning and S-scanning, respectively, are shown from the outside elliptic trajectories to the inner trajectories. As stated in last section, the performance of S-scanning scheme is better than U-scanning scheme in many aspects. And it will be the same, if the location awareness conditions are established through direction finding and emitter location capability, and are collaborated with each other among CRUs.

4. Conclusion

In this paper, we have first proposed a systematic approach with their corresponding metrics for evaluating independent and concurrent AJ and LPD performance qualitatively and quantitatively. A representative GEO satellite communication scenario for independent and concurrent AJ and LPD performance evaluations is explored thoroughly. Based on these metrics, it is direct to see that by spreading signal spectrum, complicating signal waveforms, or lowering power control uncertainty, respectively, will enhance system performance accordingly. A CRU model with U-scanning or S-scanning techniques and a quantified CPR metric for cognitive communications adapted from TRANSEC is investigated as well. In this model real-time spectrum sensing characteristics are coordinated together with system parameters in temporal and frequency domains, e.g., scanning rate and framing processing time, for evaluating the performance of CR communications under an elliptical or a hyperbolic operation scenario, which can be applied for radio spectrum sensing and location awareness in cognitive radio communications. The proposed schemes and metrics can pave one practical way for the system evaluations of cognitive radio communications.

5. References

Nicholson, D. L. (1987). Spread Spectrum Signal Design-LPE & AJ systems, *Computer science Press*, 1987

Schoolcraft, R. (1991). Low Probability of Detection Communications-Waveform Design and Detection Techniques, *IEEE Military Communications Conference*, Vol. 2, pp. 832-840, November 1991

Jain, P. C. (1990). Architectural Trends in Military Satellite Communications Systems, *Proceedings of the IEEE*, Vol. 78, no. 4, July 1990

Sklar, B. (2001). *Digital Communications*, Prentice-Hall, 2nd Ed, 2001

Turner, L. (May 1991). The Evolution of Featureless Waveforms for LPI Communications, *National Aerospace and Electronic Conference, NAECON* 1991

Glenn, A. B. (1983). Low Probability of Intercept, *IEEE Communications Magazine*, Vol. 21, pp. 26-33, 1983

Gutman, L. L. & Prescott, G. E. (December 1989). System Quality Factors for LPI Communications, *IEEE Aerospace and Electronic Systems Magazine*, Vol. 4, pp. 25-28, 1989

Dillard, G. M. & Dillard, R. A. (November 2001). A Metric for Defining Low Probability of Detection Based on Gain Differences, *IEEE Thirty-Fifth Asilomar Conference on Signals, Systems and Computers*, Vol. 2, pp. 1098-1102, November, 2001

Weeks, G. D.; Townsend, J. K. & Freebersyser, J. A. (1998). A Method and Metric for Quantitative defining Low Probability of Detection, *IEEE Military Communications Conference, MILCOM'98*, 1998

Mills, R. F. & Prescott, G. E. (1995). Waveform Design and Analysis of Frequency Hopping LPI Networks, *IEEE Military Communications Conference, MILCOM'95*, 1995

Mills, R. F. & Prescott, G. E. (July 2000). Detectability Models for Multiple Access Low Probability of Intercept Networks, *IEEE Transactions on Aerospace and Electronic Systems*, Vol. 36, (July 2000), pp. 848-858

Binia, J. (2004). LPI Communication in Channels with Absorption Loss, *IEEE Military Communications Conference, MILCOM'04*, 2004

Wu, P. H. (2005). On Sensitivity Analysis of Low Probability of Intercept (LPI) Capability, *IEEE Military Communications Conference, MILCOM'05*, 2005

Wu, P. H. (2006). Optimal Interceptor for Frequency-Hopped DPSK Waveform, *IEEE the Military Communications Conference, MILCOM'06*, 2006

Burda, K. (2004). The Performance of the Follower Jammer with a Wideband Scanning Receiver, *Journal of Electrical Engineering*, Vol. 55, (2004), pp. 36-38

Gross, F. B. & Chen, K. (April 2005). Comparison of Detectability of Traditional Pulsed and Spread Spectrum Radar Waveforms in Classic Passive Receivers, *IEEE Transactions on Aerospace and Electronic Systems*, Vol. 41, (2005), pp. 746-751

Haykin, S. (February 2005). Cognitive Radio: Brain-Empowered Wireless Communications, *IEEE Journal on selected Areas in Communication*, Vol. 23, (2005), pp. 201-220

NTIA, U.S. frequency allocations. Online available from http://www.ntia.doc.gov/osmhome/allochrt.pdf

Spectrum policy task force report. (2002). *Federal Communications Commission, Tech. Rep. 02-155*, November, 2002

Mitola III, J. & Maguire Jr., G. Q. (1999). Cognitive radio: Making software radios more personal, *IEEE Personal Communications*, Vol. 6, (1999), pp. 13-18

Mitola III, J. (May 2000). Cognitive Radio: An Integrated Agent Architecture for Software Defined Radio, *Ph. D. dissertation, Royal Institute of Technology, Sweden*, May 8, 2000

Jondral, F. K. & Karlsruhe, U. (August 2007). Cognitive Radio: A Communications Engineering View, *IEEE Wireless Communication*, (August 2007), pp. 28-33

Mody, A. N.; Blatt, S. R.; Mills,D. G. & et al. (October 2007). Recent Advances in Cognitive Communications", *IEEE Communication Magazine*, (October 2007)

Cordeiro, C.; Challapali, K.; Birru, D. & et al. (November 2005). IEEE 802.22: the First Worldwide Wireless Standard Based on Cognitive Radios, *Proceeding of 2005 First IEEE International Symposium on New Frontiers in Dynamic Spectrum Access Networks*, pp. 328-337, November 8-11, 2005

Ning, H; Hwan, S. S.; Hak, C. J. & et al. (February 2006). Spectral Correlation Based Signal Detection Method for Spectrum Sensing in IEEE 802.22 WRAN Systems, *Proceeding of 8th International Conference on Advanced Communication Technology*, Vol. 3, pp. 1765-1770, February 20-22, 2006

Cordeiro, C.; Challapali, K. & Birru, D. (April 2006). IEEE 802.22: An Introduction to the First Wireless Standard Based on Cognitive Radios, *Journal of Communicaton*, Vol. 1, No. 1, (April 2006)

Torrieri, D. J. (1986). Fundamental Limitations on Repeater Jamming of Frequency-Hopping Communications, *IEEE Journal on Selected Areas in Communication*, Vol. 7, (1986), pp. 569-575

Felstead, E. B. (October 1998). Follower Jammer Considerations for Frequency Hopped Spread Spectrum, *Proceeding of MILCOM'98*, Vol. 2, pp. 474-478, October 18-21, 1998.

Liao, C. H.; Lee, Z. S.; Tsay, M. K. & Lin, G. J. (2007). Anti-Jamming and LPD/I Performance Trade-off Analysis for Secure Communications, *International Journal of Electrical Engineering*, Vol. 14, No. 6, (2007), pp. 441-449

Liao, C. H.; Tsay, M. K. & Lee, Z. S. (February 2009). Secure Communication System through Concurrent AJ and LPD Evaluation, *Wireless Personal Communications*, Vol. 49, Issue 1, (February 2009), pp. 35-54

Zhao, Y.; Mao, S.; Neel, J.O. & et al. (April 2009). Performance Evaluation of Cognitive Radios: Metrics, Utility Functions, and Methodology, *Proceedings of the IEEE*, Vol. 97 , Issue 1, (April 2009), pp. 642-659

Jayasinghe, L. K. S.; Rajatheva, N. & Latva-aho, M. (2011), Optimal Power Allocation for PNC Relay Based Communications in Cognitive Radio, *IEEE International Conference on Communications (ICC)*, pp. 1-5, 2011

Liu, Y.; Zhao, Z. & Tang, H. (2010). Radio Resource Management between Two User Classes in Cognitive Radio Communication, *Second International Conference on Networks Security Wireless Communications and Trusted Computing (NSWCTC)*, pp. 215-218, 2010

Liang, Y. C.; Chen, K. C.; Li, G. Y. & et al. (September 2011), Cognitive Radio Networking and Communications: An Overview, *IEEE Transactions on Vehicular Technology*, Vol. 60, No. 7, (September 2011), pp. 3386-3407

6

Blind Detection, Parameters Estimation and Despreading of DS-CDMA Signals in Multirate Multiuser Cognitive Radio Systems

Crépin Nsiala Nzéza[1] and Roland Gautier[2]
[1]*SEGULA Technologies Automotive,*
Département Recherche et Innovation, Parc d'Activités Pissaloup,
[2]*Université Européenne de Bretagne,*
Université de Brest, Lab-STICC UMR CNRS 3192,
France

1. Introduction

The design of smart terminals able to detect reachable frequency bands and configure them according to the available channel state information is a major focus of current research. Two main approaches are addressed in the literature: cooperative scenario and blind methods whom transmission parameters are unknown. Blind spectrum sensing and parameters estimation (at the receiver side) may guarantee a more intensive and efficient spectrum use, a higher quality of service and more flexibility in devices self-configuration. This scenario, which can be viewed as a brick of a Cognitive Radio (CR) (Mitola, 2000) is considered throughout this chapter. We consider the more general case of Direct-Sequence (DS) spread spectrum signals in multiuser multirate CDMA systems where spreading sequences may be longer than the duration of a symbol. An easy way to view the multirate CDMA transmission is to consider the variable spreading length (VSL) technique where all users employ sequences with the same chip period. Moreover, the data rate is tied to the length of the spreading code of each user. Many researches aiming at spectrum sensing in cognitive radio context have been proposed, such as blind cyclostationary approaches (Hosseini et al., 2010), SOS approaches (Cheraghi et al., 2010), Student's t-distribution testing problem (Shen et al., 2011). Authors proved that these methods exhibited very good performances in discriminating against noise due to their robustness to the uncertainty in noise power. Nevertheless, they could be computationally complex since they may require significantly long observation time. Several blind approaches (i.e., when the process of recovering data from multiple simultaneously transmitting users without access to any training sequences) have been addressed in the literature (Buzzi et al., 2010; Khodadad, Ganji & Mohammad, 2010; Khodadad, Ganji & Safaei, 2010; Meng et al., 2010; Yu et al., 2011; Zhang et al., 2011). However, most of them require some prior knowledge about users parameters such as signature waveform, processing gain, pseudo-noise code (for a particular group of active users) or chip rate. These parameters may be unknown in a realistic non-cooperative context. Besides, a PARAFAC based method was addressed in (Kibangou & de Almeida, 2010), but in the single user case in a no noisy environment. Furthermore, in multiuser asynchronous systems both problems of blind despreading and synchronization are much more demanding, what become much more challenging. In this chapter, we addressed an iterative with deflation approach aiming to the blind multiuser multirate signals detection. Moreover, in (Koivisto & Koivunen, 2007), a similar method was proposed starting from that given in (Nsiala Nzéza, 2006), without addressing the multiuser multirate blind detection. Moreover, even though authors in (Koivisto & Koivunen, 2007), analyzing the scheme in (Nsiala Nzéza et al., 2004) claimed that good performances can be obtained in asynchronous multiuser systems, the computing time increases concomitantly with both the number of users and the correlation matrix size. Besides, the number of interfering users in (Koivisto & Koivunen, 2007) is assumed to be known at the receiver side, contrary to what is assumed in this paper. At last, similar methods to those quoted in this chapter were addressed in (Ghavami & Abolhassani, 2008; 2009). Nevertheless, authors do not improve performances, since the multiuser multirate case is was not addressed. Besides, the multiuser synchronization, even with the knowledge of certain parameters is not theoreticaly discussed. Consequently, an alternative method to those quoted above is herein proposed. This efficient and low complexity scheme does not require any prior knowledge about the transmission. The novelty of the proposed approach lies in the use of iterative algorithms combining both deflation and second-order statistical estimators methods. The application in cases of long and short spreading code transmissions is discussed and performances are investigated.

2. Signal model and assumption

Let us first consider the single user case, before dealing with both multirate and/or multiuser cases. In all cases, the uplink scenario of a DS-CDMA network is considered. This section also quickly highlights the long sequence signal model even though the application of the proposed method to long sequences case will be shown later in Section 7.

2.1 Single user signal model

2.1.1 Short spreading code case

In this case each symbol is spread by a whole spreading code, i.e., $T_s = LT_c$, where L, T_s and T_c stand for the spreading sequence gain, the symbol period and the chip period, respectively. The continuous-time single user signal at the transmitter is given by:

$$s(t) = \sum_{k=-\infty}^{+\infty} a(k)\psi(t - kT_s), \tag{1}$$

where $a(k) \notin \mathbf{a}^T = [a(0), \cdots, a(k), \cdots]$ are the transmitted real- or complex valued symbols drawn from a known constellation. $\psi(t)$ stands for the signature waveform which can be expressed as:

$$\psi(t) = \sum_{l=1}^{L} c(l)g(t - lT_c), \tag{2}$$

with $\psi(t) = 0$ for $t \notin [0, LT_c[$, and $c(l) \in \mathbf{c}^T = [c(0), \cdots, c(l), \cdots]$ being the lth chip of the spreading sequence and $g(t)$ is the impulse response of the pulse shaping filter. Recall that the same spreading code is repeated for every symbol, i.e., the system is a short-code DS-CDMA system. The signal is modulated to carrier frequency f_c, i.e., mixed with carrier

$\sqrt{2A'}e^{j(2\pi t f_c+\phi')}$, where A' and ϕ' are the power and phase. The signal is transmitted through a flat-fading channel. The received continuous-time signal can be written as:

$$y(t) = \sum_{p=1}^{P} s(t - T_p - T_d)\sqrt{2A}e^{j(2\pi t f_c+\phi)} + b(t), \tag{3}$$

where T_p is the channel propagation delay relative to the beginning of each symbol interval, T_d is the signal delay at the receiver side, $A = \mathcal{G}_p A'$ is the received power ($\mathcal{G}_p$ is the path fading factor), $\phi = \phi' - 2\pi f_c T_p$ is the phase offset and $b(t)$ is additive white Gaussian noise of variance σ_b^2, P is the number of channel taps. Equation 3 may then be expressed in a close-form as:

$$y(t) = \sum_{k=-\infty}^{+\infty} a(k)h(t - kT_s - T_d) + b(t), \tag{4}$$

where $h(t) = \sum_{p=1}^{P} \sqrt{2A}e^{j(2\pi t f_c+\phi)}\psi(t - T_p)$. Accordingly, following assumptions are made: $a(k)$ are independent, centered with variance σ_a^2; both T_p and T_d are supposed to be constant during the observation time and to satisfy $0 \leqslant T_p < T_d < T_s$; and the signal-to-noise ratio (SNR in dB) at the detector input may be negative (signal hidden in the noise).

2.1.2 Long spreading code signal model

Without loss of generality, the base-band transmission in an AWGN channel is first considered. In this case, $h(t)$ can be rewritten as $h(t) = \psi(t) = \sum_{l=1}^{L} c(l)g(t - lT_c)$. Moreover, in a long spreading code approach, a whole code spread Q_s ($Q_s \in \mathbb{N}^*$) consecutive symbols. Therefore, each symbol can be viewed as spread by a code of length $L_s = \frac{L}{Q_s}$ (i.e., $T_s = L_s T_c$). Let us also define $\psi_s(t)$ as $\psi_s(t) = 0$ for $t \notin [0, L_s T_c[$. Hence, Equation 4 becomes:

$$y(t) = \sum_{k=-\infty}^{+\infty} a(k)\psi_s(t - kT_s - T_d) + b(t), \tag{5}$$

where $\psi_s(t) = \sum_{l=(\langle k \rangle_{Q_s})L_s}^{(\langle k+1 \rangle_{Q_s})L_s - 1} c(l)g_s(t - lT_c)$, $\langle x \rangle \equiv x \textit{ modulo } Q_s$. First, setting $L_s = L$ and $Q_s = 1$ leads to the short spreading code case, i.e.,

$$y(t) = \sum_{k=-\infty}^{+\infty} a(k)\psi(t - kT_s - T_d) + b(t). \tag{6}$$

Secondly, by setting $h_s(t) = \sum_{p=1}^{P} \sqrt{2A}e^{j(2\pi t f_c+\phi)}\psi_s(t - T_p)$, Equation 5 can be expressed as Equation 4, taking into account P channels taps as:

$$y(t) = \sum_{k=-\infty}^{+\infty} a(k)h_s(t - kT_s - T_d) + b(t). \tag{7}$$

Furthermore, setting $L_s = L$, Equation 7 leads to a similar expression than in Equation 4. Therefore, in the sequel, only the short spreading code will be considered for multiuser signal model. However, when the long spreading code case will be discussed, there will be a special indication.

2.2 Multiuser multirate signal model

The VSL-based asynchronous DS-CDMA system where $\{R_0, R_1, \cdots, R_{S-1}\}$ stands for a set of S available data rates is considered. A different slow fading multipath channel with i.i.d Rayleigh random variables and unity second moment fading amplitudes is assumed for each interfering user. Let us set N_u^i the number of active users transmitting at R_i (hence, belonging to group i) and N_u the total number of users such that $\sum_{i=0}^{S-1} N_u^i = N_u$. The nth transmitted signal at R_i, denoted throughout this chapter as the $(n,i)^{th}$ user, is written as:

$$s_{n,i}(t) = \sum_{k=-\infty}^{+\infty} a_{n,i}(k)\psi_{n,i}(t - kT_{s,i}), \tag{8}$$

where $\psi_{n,i}(t) = \sum_{l=1}^{L_i} c_{n,i}(l)g_i(t - lT_c)$ represents the $(n,i)^{th}$ user's signature code, i.e., the convolution of the transmission filter with the spreading sequence $\{c_{n,i}(l)\}_{l=1\cdots L_i}\}$ chipping pulse. L_i is the spreading factor for the $(n,i)^{th}$ user or for the nth transmitted signal at R_i,i.e., with the symbol rate $T_{s,i}$. Moreover, let us assume that the channel seen by users in group i is different from that seen by ones in another group, but with the same number P of paths. $\mathcal{G}_{p,i}$ and $T_{p,i}$ represent the pth path fading factor and its corresponding transmission delay which typically satifies: $0 \leq T_{p,i} << T_{d_{n,i}} < T_{s_i}$. The $(n,i)^{th}$ user channel corrupted received signal is done by:

$$y_{n,i}(t) = \sum_{k=-\infty}^{+\infty} a_{n,i}(k)h_{n,i}(t - kT_{s,i} - T_{d_{n,i}}) + b(t), \tag{9}$$

where $h_{n,i}(t) = \sum_{p=0}^{P-1} \sqrt{2A_i}e^{j(2\pi t f_c + \phi_i)}\psi_{n,i}(t - T_{p,i})$, A_i and ϕ_i are defined as for Equation 3 using $\mathcal{G}_{p,i}$, $T_{d_{n,i}}$ stands for the $(n,i)^{th}$ user's delay at the reciver side, which satifies: $T_{d_{n,i}} < T_{s_i}$. In Equation 9, it was assumed that users in group i are transmitted with the same power in order to further simplify theoretical analysis. However it is not required in practise with respect to iterative implementation discussed in Section 6. Concluding for Equation 9, the channel corrupted multiuser signal may be modelled as:

$$y(t) = \sum_{i=0}^{S-1} \sum_{n=0}^{N_u^i - 1} \sum_{k=-\infty}^{+\infty} a_{n,i}(k)h_{n,i}(t - kT_{s_i} - T_{d_{n,i}}) + b(t). \tag{10}$$

Also, using Equations 5 and 10, the long code multiuser multirate signal model may be written as:

$$y(t) = \sum_{i=0}^{S-1} \sum_{n=0}^{N_u^i - 1} \sum_{k=-\infty}^{+\infty} a_{n,i}(k)h_{s_{n,i}}(t - kT_{s_i} - T_{d_{n,i}}) + b(t). \tag{11}$$

3. Description of the proposed approach

Fig. 1 shows the different steps of the proposed approach in a sequential manner for getting the full picture of the subject. Since we focus on the last three steps in this chapter, the spectrum sensing step may be considered as a preliminary one, hence it will be assumed in what follows that this step has been performed. Even so, the reader will find details in (Nzéza et al., 2009) about the proposed method for this purpose. Spectral components (e.g., central frequency bandwidth) are estimated by the averaged periodogram non-parametric approach using a Fast Fourier Transform (FFT) combined with a detection threshold. This

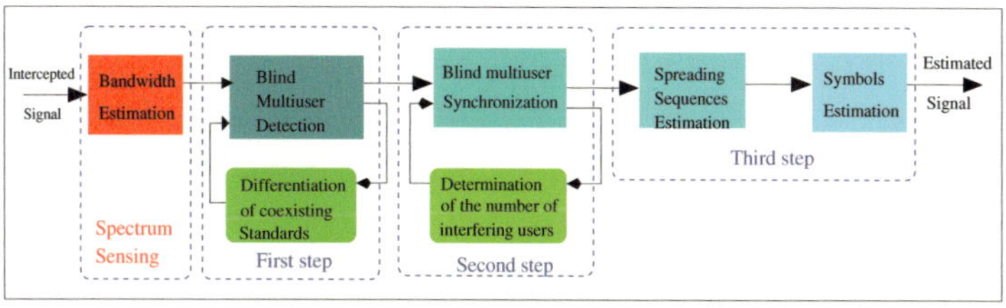

Fig. 1. Illustration of different stages of the proposed method (after the bandwidth estimation process).

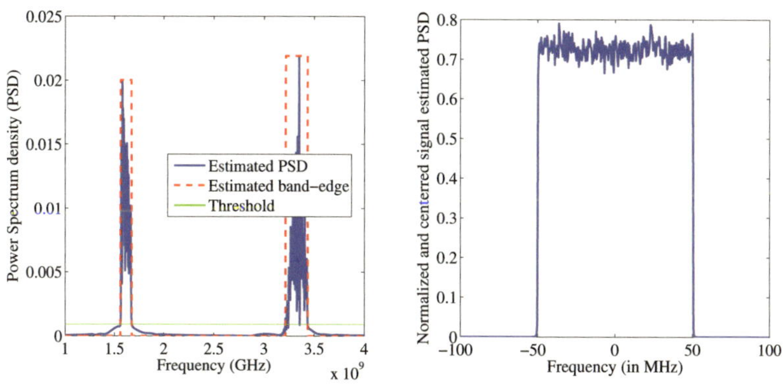

(a) Filtered periodogram with the estimated threshold

(b) Estimaton of the bandwidth @1.62 GHz

Fig. 2. Blind bandwidth estimation and center frequencies recovering.

threshold is computed from the signal received power estimation as shown on Fig. 2 (a), obtained with following parameters: One signal at 1.62 *GHz* and another at 3.3 *GHz*, spread with complex GOLD sequences with L=127, T_c=150 *MHz*, F_e=300 *MHz*, T_F=2μs, K=300, *QPSK* modulation, COST207RAx6 (Committee, 1989) channel, and the SNR = − 3 dB at the receiver side. If we are interested by the signal at 1.62 *GHz*, we deduct a frequency-band of $\widetilde{W}$=100 *MHz*, as evidenced on Fig. 2 (b). Note that if two or more signals share the same bandwidth, their differenciation is performed through the blind detection step, which is the first one on Fig. 1.

The first steps allows to detect signals and to estimate their data rate through the analysis of fluctuations of correlation estimators (Nsiala Nzéza, 2006) whithin each identified bandwidth of interest. Besides, it allows a multi-standard detection through a differentiation of various standards data rate as suggested in (Williams et al., 2004). Successive investigations of the contributions of noise and channel-corrupted signal through the analysis of the second-order moment of the correlation estimator computed from many randomly-selected analysis windows constitutes the key point of the proposed temporal parameters estimation approach. As a result, we compute a function which is a measurement of these fluctuations. The obtained curve highlights equispaced peaks of different amplitudes, therefore for different symbol periods which leads to a low computational complexity and an efficient estimation of symbols duration. The proposed scheme is also insensitive to phase and frequency offsets since it is the square modulus which is computed.

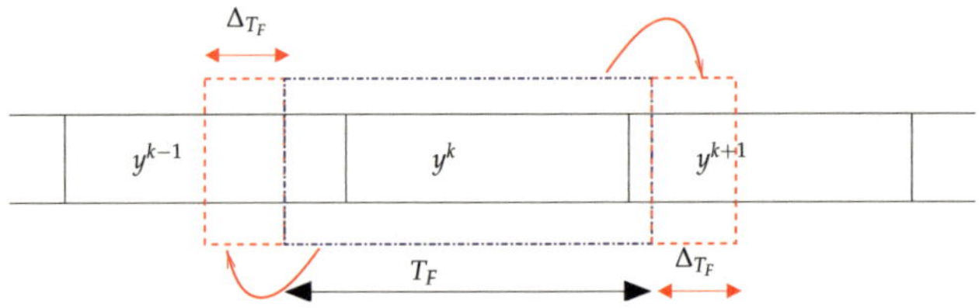

Fig. 3. Correlation computation over an analysis window.

The knowledge of symbol durations permits to perform the synchronization process. This constitutes the second step as highlighted on Fig. 1. Moreover, through the analysis of correlation matrix at this stage, the synchronization may be done iteratively by deflation using, as appropriate, one of the criteria detailed in Sections 4 and 5. This process also allows to determine the number of interfering users transmitting at the same data rate by determinating the number of the synchronization peaks. Finally, in the third step on Fig. 1, codes and symbols are recovered even at very low SNR using linear algebra techniques, also described in Sections 4 and 5. The assessment of those parameters constitues a brick of a CR's receiver and allows its self-reconfigurability. It is very important to note that although the overall method is presented sequentially, it must be implemented iteratively, as discussed in Section 6, and further applied to long spreading codes case in Section 7.

4. Application to short spreading codes in the single user case

4.1 Estimation of the symbol duration

The analysis of the autocorrelation fluctuations estimators allows to achieve a direct blind signals detection and standards differentiation. The estimation process of the symbol duration T_s uses correlation properties of the received signal. Using K temporal windows of duration T_F, an autocorrelation estimation $\widehat{R}_{yy}^k$ of the received signal can be written as:

$$\widehat{R}_{yy}^k(\theta) = \frac{1}{T_F + 2\Delta_{T_F}} \int_{-\Delta_{T_F}}^{T_F+\Delta_{T_F}} y^k(t)(y^k)^*(t-\theta)dt, \tag{12}$$

where $y^k(t)$ is the signal sample over the kth window and $(\cdot)^*$ denotes the conjugate transpose of $(\cdot)$. In order to avoid edge effects, $\widehat{R}_{yy}^k$ is computed within each window k and also during Δ_{T_F} at both right and left side of the kth window, as illustrated on Fig. 3. Equation 12 constitutes the main key of the detection algorithm, since it reduces constraints on the window duration T_F, and thus on the total number K of analysis windows. Δ_{T_F} can be theoretically neglicted compared to T_F ($\frac{T_F}{4} \leq \Delta_{T_F} \leq \frac{T_F}{2}$), since Equation 12 is computed from many analysis windows. However, it is necessary to calculate it in this manner in order to avoid edge effects. Without loss of generality, Equation 12 can be reexpressed more simply for further theoretical analysis as:

$$\widehat{R}_{yy}^k(\theta) = \frac{1}{T_F} \int_{0}^{T_F} y^k(t)(y^k)^*(t-\theta)dt. \tag{13}$$

Moreover, from simulations results, which will be discussed later, the condition $\frac{T_F}{4} \leq \Delta_{T_F} \leq \frac{T_F}{4}$ seems to be a good choice of Δ_{T_F}. $\widehat{R}_{yy}(\theta)$ is computed over K windows, and its second-order moment is given by:

$$\widehat{\mathcal{E}}\left\{|\widehat{R}_{yy}(\theta)|^2\right\} = \frac{1}{K}\sum_{k=0}^{K-1}|\widehat{R}^k_{yy}(\theta)|^2 = \Phi(\theta), \tag{14}$$

where $\widehat{\mathcal{E}}(\cdot)$ is the estimated expectation of $(\cdot)$. Hence, $\Phi(\theta)$ is a measurement of the fluctuations of $\widehat{R}_{yy}(\theta)$, as proved in the sequel. Note that it is assumed an average $\mu_{\widehat{R}_{yy}}(\theta)=\mathcal{E}\left\{|\widehat{R}_{yy}(\theta)|\right\}=0$ in Equation 14 for theoretical analysis. In practise, $\Phi(\theta)$ may be computed as:

$$\Phi(\theta) = \widehat{\mathcal{E}}\left\{|\widehat{R}_{yy}(\theta)|^2 - |\mu_{\widehat{R}_{yy}}(\theta)|^2\right\}. \tag{15}$$

Equation 15 represents the centered second order moment of the magnitude of the correlation fluctuations or variance. However, since fluctuations are computed from many randomly-selected windows, Equation 14 is more suitable for theoretical purpose. The difference between both Equations 15 and 14 lies in the magnitude of fluctuations peaks. With assumptions made in Section 2, we have:

$$\widehat{R}_{yy}(\theta) \simeq \widehat{R}_{ss}(\theta) + \widehat{R}_{bb}(\theta), \tag{16}$$

where $\widehat{R}_{ss}(\theta)$ and $\widehat{R}_{bb}(\theta)$ are the estimates of the noise-free signal $s(t)$ and that of the noise autocorrelation fluctuations, respectively. Since symbols are assumed to be independent, uncorrelated with the noise, the variance of $\widehat{R}_{yy}(\theta)$ using Equation 14 is done by:

$$\Phi(\theta) = \Phi_s(\theta) + \Phi_b(\theta), \tag{17}$$

where $\Phi_s(\theta)$ and $\Phi_b(\theta)$ stand for the fluctuations due to the noise-free signal and that due to the additive random-noise, respectively. Expression 17 proves that $\Phi(\theta)$ is a measurement of the variations of the estimator of autocorrelation fluctuations. Since fluctuations are computed from many randomly-selected windows, they do not depend on the signal relative delay, nor on channel path delays. Channel gains act as a multiplicative factor in the fluctuations curve, as it will be shown hereinafter.

4.1.1 Noise contribution to global fluctuations $\Phi(\theta)$

Assume a receiver filter with a flat frequency response in $[-W/2, +W/2]$ and null outside. As proved in (Nsiala Nzéza, 2006), fluctuations $\Phi_b(\theta)$ are uniformly distributed over all values of θ. Hence, they can be characterized by their mean m_{Φ_b} and standard deviation σ_{Φ_b} as:

$$m_{\Phi_b} = \frac{\sigma_b^4}{WT_F} \quad \text{(a)}, \quad \sigma_{\Phi_b} = \sqrt{\frac{2}{K}}\frac{\sigma_b^4}{WT_F} \quad \text{(b)}. \tag{18}$$

Equation 18 (b) shows that increasing the number K of windows improves the detection by lowering noise contribution.

4.1.2 Signal contribution to global fluctuations $\Phi(\theta)$

By only considering the noise-free signal, we demonstrate in (Nsiala Nzéza, 2006) that on average, high amplitudes of its fluctuations, set to as $\Phi_s(\theta)$, occur for θ multiple of T_s. Then,

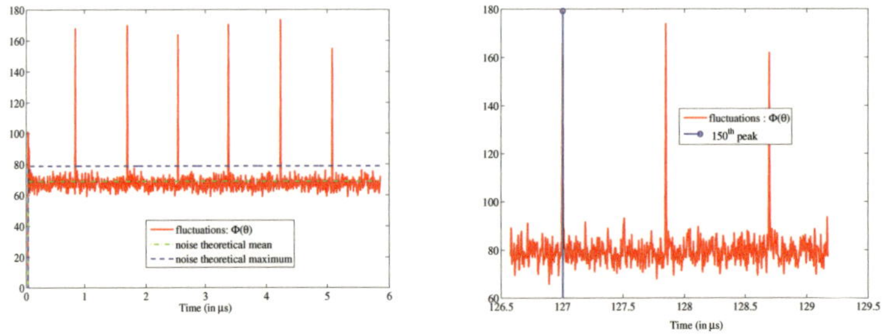

(a) Fluctuations $\Phi(\theta)$ (b) Estimation of $\widetilde{T}_s$

Fig. 4. Estimation of the symbol period through fluctuations $\Phi(\theta)$ analysis , SNR $= -7$ dB in uplink.

for each θ multiple of T_s, both fluctuations average amplitude m_{Φ_s} and standard deviation σ_{Φ_s} are given by:

$$m_{\Phi_s} = \frac{T_s}{T_F}\widetilde{\sigma}_s^4 = \frac{LT_c}{T_F}\widetilde{\sigma}_s^4, \text{ and } \sigma_{\Phi_s} = \sqrt{\frac{2}{K}}m_{\Phi_s} = \sqrt{\frac{2}{K}}\frac{T_s}{T_F}\widetilde{\sigma}_s^4, \tag{19}$$

where $\widetilde{\sigma}_s^2$ stands for the channel-corrupted signal power at the receiver side. Equation 19 shows that the fluctuations curve will exhibit equispaced peaks which average spacing correspond to the estimated symbol period, which allows at itself to do standards differentiation, as illustrated in Fig. 4.

Fig. 4 was obtained with following simulations parameters: Complex GOLD sequence with L=127, T_c=100 MHz, F_e=300 MHz, T_F=2μs, K=300, $QPSK$ modulation, COST207RAx6 Committee (1989) channel, and the SNR $= -7$ dB at the receiver side. Equation 19 also indicates that the average fluctuations amplitude is tied to the sequence length and the signal power at the receiver side. Continuing with Equation 18, a theoretical detection threshold is taken as: $\zeta = m_{\Phi_b} + 3 \cdot \sigma_{\Phi_b}$. This equation shows that, whenever a spread spectrum signal is hidden in the noise, the average of the curve deviates from the theoretical average. Especially, the curve maximum is above the noise fluctuations theoretical maximum. It is precisely this curve behaviour which allows hidden signals detection. The reader may find extra detailed theoretical performance analysis of this blind detection scheme in (Nzéza et al., 2008).

4.2 Synchronization process analysis

At this point, only T_s is known. The signal is sampled and divided into N non-overlapping temporal windows of duration $T_F = T_s = MT_e$, $M \in \mathbb{N}^\star$, where T_e is the sampling period. Thus, each window contains M samples. Sampling and chip periods are not equals, and the number of samples per window is not equal to the code length, since these parameters are unknown. From the sampled-received signal vector $\mathbf{y}^e(t)$=$[s(t), s(t+T_e), \cdots, s(t+(M-1)T_e)]$, the $(M \times N)$-matrix $\mathbf{Y}^e$, which N columns contain M signal samples is computed as:

$$\mathbf{Y}^e = \begin{pmatrix} s(t) & \cdots & s(t+(N-1)T_s) \\ \vdots & \cdots & \vdots \\ s(t+T_s-T_e) & \cdots & s(t+NT_s-T_e) \end{pmatrix}. \tag{20}$$

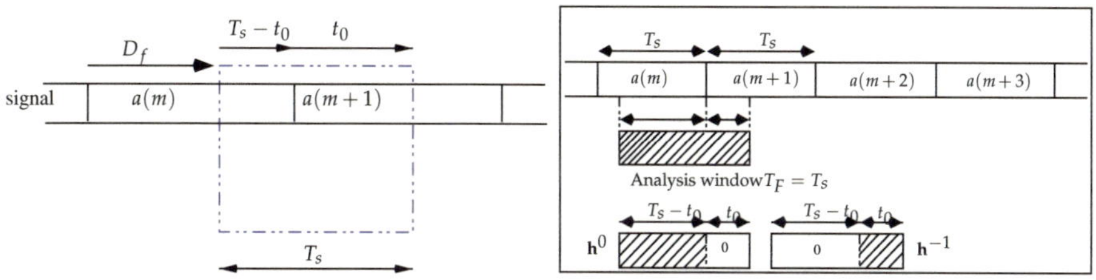

(a) Windows position influence (b) Vectors $\mathbf{h}^0$ and $\mathbf{h}^{-1}$ computation

Fig. 5. Relative position of signals and an analysis window before the synchronization process in uplink.

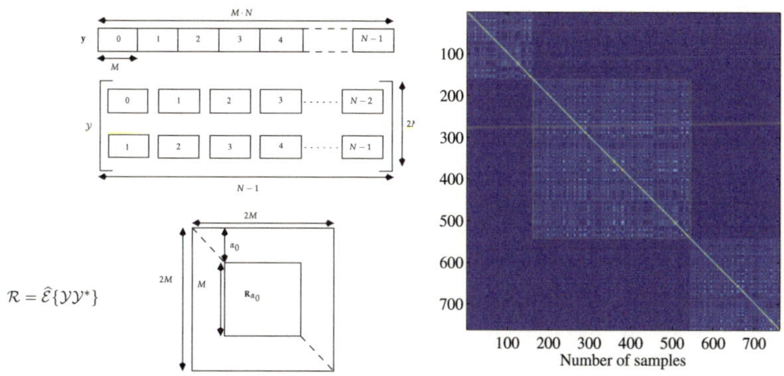

(a) Vector $\mathcal{Y}$ and matrix $\mathcal{R}$ computation (b) Computation of the matrix $\mathcal{R}$

Fig. 6. Double-size correlation matrix estimation with $\widetilde{T_s}$, SNR $= -7$ dB in uplink.

Let us analyze Fig. 5 (a) where D_f and t_0 stand for analysis window shifts (here, from left towards right) and the temporal shift between the analysis window and the beginning of a whole symbol, respectively. It clearly appears that any shift D_f induces t_0 changes, as well as in Equation 20. As proved in (Nsiala Nzéza, 2006), since the filter $h(t)$ is defined in $[0 \quad T_s[$, Equation 20 can be rewritten as:

$$\mathbf{Y}^e = \left(\mathbf{h}^0 + \mathbf{h}^{-1}\right)\mathbf{a}^T + \mathbf{b}^e, \tag{21}$$

where vector $\mathbf{a}^T = [\cdots, a(m)\cdots]$ contains all symbols, and vectors $\mathbf{h}^0$ and $\mathbf{h}^{-1}$ are defined as follows: $\mathbf{h}^{-1}$ contains the end of the corresponding spreading waveform convolved with the channel during $T_s - t_0$, followed by zeros during t_0, meanwhile $\mathbf{h}^0$ contains zeros during t_0, followed by the beginning of the corresponding spreading waveform convolved with the channel during $T_s - t_0$, as illustrated on Fig. 5 (b). In Equation 21, $\mathbf{b}^e$ stands for the noise $(M \times N)$-matrix, the received signal $(M \times M)$ correlation matrix $\mathbf{R}$ may be computed as : $\mathbf{R} = \widehat{\mathcal{E}}\{\mathbf{Y}^e(\mathbf{Y}^e)^*\}$. In practise, in order to reduce the computational time, we compute the double size matrix $\mathcal{R} \in \mathbb{C}^{2M\times 2M}$ containing the matrix $\mathbf{R}$ induced by the shift t_0, as shown on Fig. 6 (a). Fig. 6 (b) highlights the matrix $\mathcal{R}$ in simulation with the same parameters than on Fig. 4. Moreover, let us assume the signal energy ($\backsimeq T_e\|\mathbf{h}|^2$) uniformly distributed over a period symbol; we obtain the following approximation which is statistically valid if the code length is large enough (practically it is):

$$\|\mathbf{h}^0\|^2 \backsimeq (1-\alpha_0)\|\mathbf{h}\|^2, \quad \text{and} \quad \|\mathbf{h}^{-1}\|^2 \backsimeq \alpha_0\|\mathbf{h}\|^2, \tag{22}$$

where $\alpha_0 = \frac{t_0}{T_s}$. Hence, the estimated correlation matrix can be written as:

$$\mathbf{R} = \sigma_b^2 \left\{ \beta \left\{ (1-\alpha_0)\mathbf{v}^0(\mathbf{v}^0)^* + \alpha_0 \mathbf{v}^{-1}(\mathbf{v}^{-1})^* \right\} + \mathbf{I} \right\}, \tag{23}$$

where $\beta = \rho\frac{T_s}{T_e}$, ρ stands for the signal-to-noise and interference ratio (SNIR), $\mathbf{v}^0$ and $\mathbf{v}^{-1}$ are normalized vectors of $\mathbf{h}^0$ and $\mathbf{h}^{-1}$, and $\mathbf{I}$ is the identity matrix. From Equation 23, an eigenvalue decomposition highlights 2 eigenvalues associated to the signal space and $M-2$ others associated to the noise space (all are assumed to be equal on average to the noise power). Since sequences are supposed uncorrelated, the eigenvalues in an unspecified order are:

$$\lambda_1^0 = \sigma_b^2\{\beta(1-\alpha_0)+1\}, \quad \lambda_2^{-1} = \sigma_b^2\{\beta\alpha_0+1\}, \quad \lambda_m = \sigma_b^2, \quad m = 2, \cdots, M. \tag{24}$$

The synchronization consists in adjusting more precisely the symbol period by estimating the beginning of the first whole symbol. It is obvious to check that the eigenvalues sum is constant (with a concentration around certain values), and especially does not depend on α_0. Therefore, the suitable criterion in order to highlight the phenomenon of concentration is the sum of squares. The FROBENIUS square norm (FSN) of Equation 23 is defined as the sum of the square eigenvalues of Equation 24. Then, after simplifications, we get:

$$\|\mathbf{R}\|^2 = \sigma_b^4\left\{ \left(1+2\beta-\beta^2\right) + M \right\} + 2\beta^2\sigma_b^4\left\{1-\alpha_0+\alpha_0^2\right\}. \tag{25}$$

Simplified Equation 25 proves that the sum of square eigenvalues is not sensitive to neither transmission delays nor shifts. The FSN-based (FSNB) criterion is defined as being the constant part of Equation 25, set to F:

$$F(\alpha_0) = 1 + \left(\alpha_0^2 - \alpha_0\right). \tag{26}$$

Maximizing the FSN of $\mathbf{R}$ is equivalent in determining the maxima of F (synchronization peak). From Equation 26, F is a convex quadratic function of α_0 over $[0, 1[$, since there is a periodicity on relative normalized positions $d_f = \frac{D_f}{T_s}$ of an analysis window and that of the signal as highlighted in Fig. 5 (a) and demonstrated in (Nsiala Nzéza, 2006)[pp. 88-115]. By setting $\langle x\rangle \equiv x \textit{ modulo } 1, x \in \mathbb{R}, \tau = \frac{T_d}{T_s}$ leads to $\alpha_0 = \langle d_f - \tau\rangle$. Consequently, since the delay τ is assumed constant, Equation 26 only depends on shifts d_f:

$$F(d_f) = 1 + \{\langle d_f - \tau\rangle^2 - \langle d_f - \tau\rangle\}. \tag{27}$$

Resulting in a maxima of F obtained for $d_f = \tau \Longrightarrow \alpha_0 = 0$, as shown on Fig. 7 (a). Once the signal is synchronized, the sequence identification process can be performed with the extracted matrix $\mathbf{R}_{\alpha_0}$, as illustrated in Fig. 7 (b).

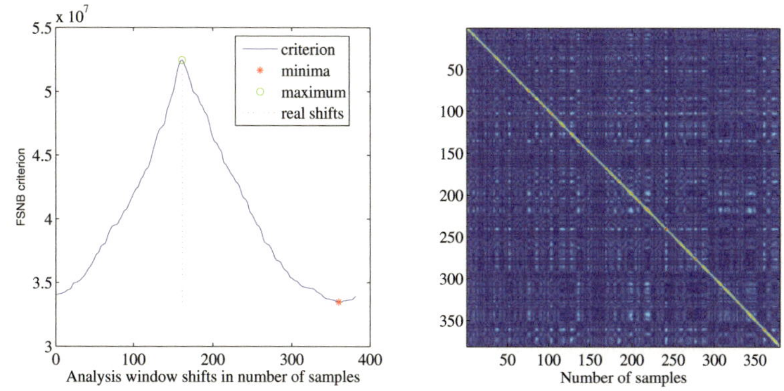

(a) FSNB criterion F (b) Extracted matrix $\mathbf{R}_{\alpha_0}$

Fig. 7. Synchronization criterion and extracted matrix $\mathbf{R}_{\alpha_0}$, SNR $= -7$ dB in uplink.

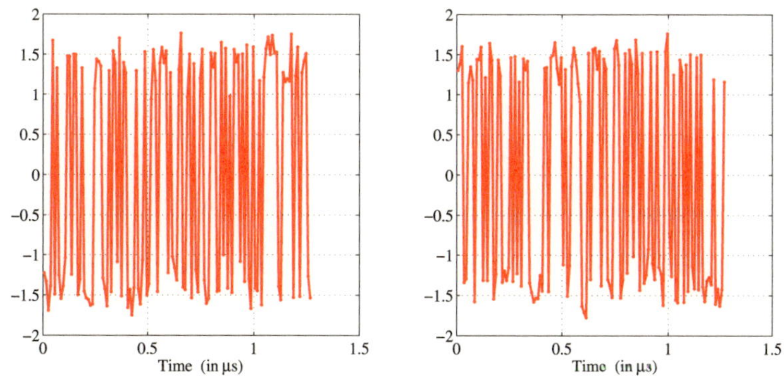

(a) Real part of the code (b) Imaginary part of the code

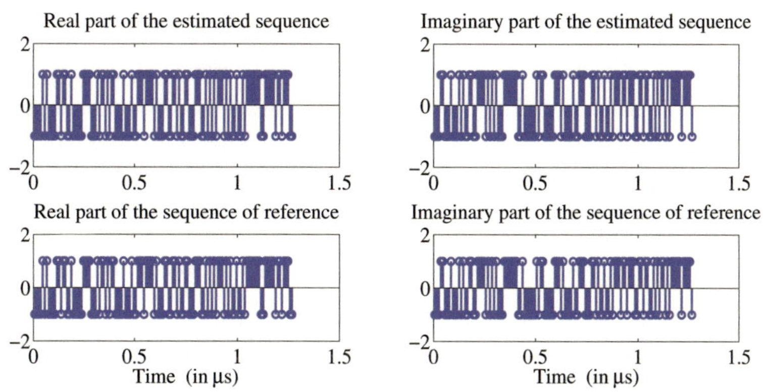

(c) Real binary part of the code (d) Imaginary binary part of the code

Fig. 8. Spreading sequence estimation and *binarization*, SNR $= -7$ dB in uplink.

4.3 Spreading code and symbols estimation

The synchronization process leads to set α_0=0, therefore, the correlation matrix becomes:

$$\mathbf{R}_{\alpha_0} = \sigma_b^2 \left\{ \beta \mathbf{v}_{\alpha_0} \mathbf{v}_{\alpha_0}^* + \mathbf{I} \right\}. \quad (28)$$

Equation 28 highlights a maximum eigenvalue which associated eigenvector contains the corresponding spreading sequence (excluding the effects of the global transmission filter), and $M-1$ eigenvalues (equal on average to the noise power at the receiver side). The largest eigenvalue corresponds to the eigenvector $\mathbf{v}_{\alpha_0}$, i.e., the signal spreading sequence (up to the channel effects).

The eigenvectors are calculated up to a complex multiplicative factor due to diagonalization. The complex factor effects are cancelled by normalizing the estimated eigenvector phase. It is made by maximizing its real part and imposing the positivity of its first component real part. In order to determine if the sequence whether initially complex or real, the variances of its real and imaginary parts (here $\mathbf{v}_{\alpha_0}$) after normalization are compared. The sequence is real when the standard deviation of its real part-to-the standard deviation of its imaginary part ration exceeds 2.5. This threshold seems to be sufficient because of eigenvectors normalization.

This stage allows us to eliminate the effects of the global transmission filter, as shown on Fig. 8 (a) and (b). The last stage is the *binarization* detailed in (Nsiala Nzéza, 2006). It consists in determining the chip period T_c (hence the length of the sequences) before seeking the binary sequence the nearest to the spreading sequence estimated in the least squares sense for each of the sequences to be estimated, as highlighted on Fig. 8 (c) and (d).

5. Extension to the multiuser/multirate case

Starting with Equation 11, we will progressively extend to the multiuser/multirate case the blind methods developped in Section 4. The assumptions of independent, centered and noise-uncorrelated signals leads to the following equations:

$$\widehat{R}_{ss}(\theta) = \sum_{i=0}^{S-1} \sum_{n=0}^{N_u^i-1} \widehat{R}_{s_{n,i}s_{n,i}}(\theta), \text{ and } \widehat{R}_{yy}(\theta) = \widehat{R}_{ss}(\theta) + \widehat{R}_{bb}(\theta), \quad (29)$$

where $\widehat{R}_{s_{n,i}s_{n,i}}(\theta)$ and $\widehat{R}_{bb}(\theta)$ are the estimates of the $(n,i)^{th}$ noise-unaffected signal and that of the noise autocorrelation fluctuations, respectively. Indeed, since the fluctuations are computed from many randomly-selected windows, they do not depend on the signals relative delays $T_{d_{n,i}}$ nor on channel paths delays. The Multiple Access Interference (MAI) noise impact is similar to that of additive noise. So, as in single-user case, both MAI and Gaussian Additive noise fluctuations are uniformly distributed over the desired frequency band. Moreover, channel gains act as a multiplicative factor on fluctuations average amplitude.

Let us consider the channel-corrupted signal and the assumptions made in Section 2. Therefore, it is fairly easy to demonstrate that on average, high amplitudes of the fluctuations of the autocorrelation estimator, denoted $\Phi_s(\theta)$, are obtained for each θ multiple of each symbol period T_{s_i}, $i = 0, \cdots, S-1$. Since the symbol periods are different, let us denote $\Phi_i(\theta)$ the fluctuations of the autocorrelation estimator of the N_u^i signals $s_{n,i}(t)$ transmitting at the same data rate T_{s_i}. Let us also term m_{Φ_i} the mean value of the fluctuations $\Phi_i(\theta)$ for each value θ multiple of T_{s_i}; it leads to:

$$\Phi_i(\theta) = m_{\Phi_i} \cdot pgn_{T_{s_i}}(\theta), \tag{30}$$

where $pgn_{T_{s_i}}(\theta) = \sum_{k=-\infty}^{+\infty} \delta(\theta - kT_{s_i})$, and the function $\delta(\theta)$ is the KRONECKER function. Consequently, the signals being independent and centered by assumption, the fluctuations $\Phi_S(\theta)$ of the global noise-unaffected signal, for each value θ multiple of T_{s_i}, can be expressed as:

$$\Phi_S(\theta) = \sum_{i=0}^{S-1} \Phi_i(\theta) = \sum_{i=0}^{S-1} m_{\Phi_i} \cdot pgn_{T_{s_i}}(\theta). \tag{31}$$

Thus, theoretical calculations developed in Subsection 4.1.2 remain valid in this case, considering each group i separately. Hence, the average amplitude of the fluctuations in each group i becomes:

$$m_{\Phi_i} = \sum_{n=0}^{N_u^i-1} \frac{T_{S_i}}{T_F} \sigma_{S_{n,i}}^4 = N_u^i \frac{T_{S_i}}{T_F} \sigma_{S_i}^4 = N_u^i \frac{L_i T_C}{T_F} \sigma_{S_i}^4, \tag{32}$$

where $\sigma_{S_i}^4 = \frac{1}{N_u^i} \sum_{n=0}^{N_u^i-1} \sigma_{S_{n,i}}^4$ can be considered as the average received power of signals within the group i. Equation 32 evidences that, the higher the number of users in a group i is, the greater the average amplitude of the fluctuations due to signals within this group is. Then, let us set to $m_{\Phi_{snir}}$ both MAI and Gaussian Additive noise contribution to fluctuations. Therefore, the longer the sequence is, the higher the peaks of correlation fluctuations are, and the biggest amplitude is usually exhibited by the fluctuations of the users transmitting at the lowest data rate. This approach is a powerful tool to estimate symbol periods, and it allows to differentiate the various transmitted data rate and to distinguish between the different standards. As proved in (Nsiala Nzéza, 2006), their mean and standard deviation are given by:

$$m_{\Phi_{snir}} = \frac{\sigma_{snir}^4}{WT_F} \quad \text{(a)},$$
$$\sigma_{\Phi_{snir}} = \sqrt{\frac{2}{M} \frac{\sigma_{snir}^4}{WT_F}} \quad \text{(b)}, \tag{33}$$

where σ_{snir}^4 represents both MAI and Gaussian Additive noise power. Completing, using Equations 30, 31, 32 and 33, the global fluctuations can be expressed as follows:

$$\Phi(\theta) = \Phi_S(\theta) + \Phi_{snir}(\theta). \tag{34}$$

Equation 34 shows that the overall fluctuations are composed of fluctuations due to the signal and noise (Gaussaian additive and MAI). It also proves that only the contribution of the signal exhibits peaks at multiple of T_{s_i} in each group i. The global fluctuactions curve highlights a superposition of regularly spaced peaks within each group i, while the noise is uniformly distributed in the band. As a result, within each group of i, the blind synchronization process can be performed, also allowing to determine the number of interfering users N_u^i as detailed in Subsection . In addition, note that the impact of MAI noise becomes negligible if the interest first goes to peaks with the highest amplitude fluctuations (Nzéza et al., 2006), i.e., to signals transmitted at $\max_{i=0,\cdots,S-1} \{T_{s_i}\}$.

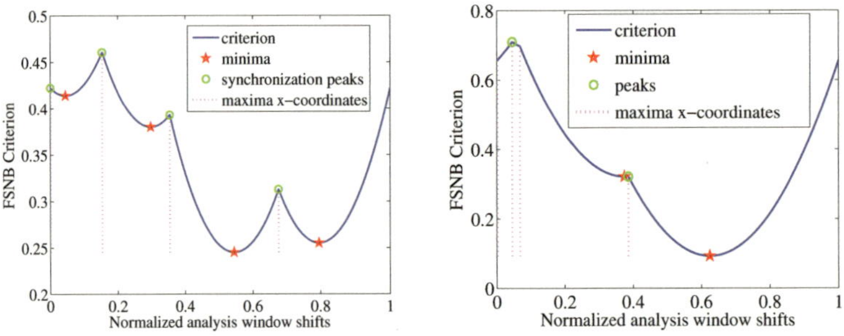

(a) ideal case

(b) synchronization peaks masking

Fig. 9. Theoretical FSNB criterion F in uplink for two cases.

5.1 Blind multiuser synchronization process

We first extend FSNB criterion to show that this criterion may exhibit the phenomenon of synchronisation peaks masking. Therefore, the modifed Pastd algorithm (Nsiala Nzéza, 2006)[pp. 115-116] may be jointly performed. Alternatively, we present another criterion based on the Maximum Eigenvalue Behaviour (MEVB) according to analysis window shifts. MEVB-based criterion provides a significant improvement of performances which is mainly due to suppression of synchronisation peaks masking that occurred with the FSNB criterion.

5.1.1 FSNB criterion theoretical analysis

For more clearness, let us set $T_{s_i}=T_s, (*)_{n,i}=(*)_n$ within a group i. Since normalized shifts α_n restrict the study into the interval $[0, \quad 1[$, there is a periodicity on the relative positions of an analysis window and that of signals. By setting $\langle x \rangle \equiv x \; modulo \; 1$, $x \in \mathbb{R}$, $\tau_n = \frac{T_{d_n}}{T_s}$ and $d_f = \frac{D_f}{T_s}$, leads to $\alpha'_n = \langle d_f - \tau_n \rangle$, $n = 0, \cdots, N_u^i - 1$. Consequently, since delays τ_n are assumed constant, Equation 27 which only depends on shifts d_f becomes:

$$F(d_f) = 1 + \sum_{n=0}^{N_u^i - 1} \{ \langle d_f - \tau_n \rangle^2 - \langle d_f - \tau_n \rangle \}. \tag{35}$$

Equation 35 shows that each signal is synchronized when d_f is equal to its corresponding transmission delay. Let us recall that a peak is a point of a curve from which, while moving by lower or higher values, the curve is always decreasing. As well as synchronization peaks are expected at points $d_f = \tau_n$. This implies to examine criterion F behaviour within $]\tau_{n-1}, \quad \tau_n]$ and $[\tau_n, \quad \tau_{n+1}[$. However for $n = 0$, or for $n = N_u^i - 1$, intervals $]\tau_{-1}, \quad \tau_0]$ and $[\tau_{N_u^i - 1}, \quad \tau_{N_u^i}[$ do not exist. Thus, by setting $\tilde{\tau}_n = \tau_{\langle n \rangle_{N_u^i}}, \quad \langle n \rangle_{N_u^i} \equiv n \; modulo \; N_u^i$ leads to take these intervals into account, and Equation 35 can be rewritten as:

$$F(d_f) = 1 + \sum_{n=0}^{N_u^i - 1} \{ \langle d_f - \tilde{\tau}_n \rangle^2 - \langle d_f - \tilde{\tau}_n \rangle \}. \tag{36}$$

Equation 36 is a convex quadratic function of d_f over any interval $[\tilde{\tau}_n, \quad \tilde{\tau}_{n+1}[$ whose peaks should be located at points such that $d_f = \tilde{\tau}_n$, $n = 0, \cdots, N_u^i - 1$, as illustrated in Fig. 9 (a), where $N_u^i = 4$, $\tilde{\tau}_0 = 0$, $\tilde{\tau}_1 = 0.1509$, $\tilde{\tau}_2 = 0.3784$ and $\tilde{\tau}_3 = 0.6979$. However, from Equation 36, we demonstrated that according to $\tilde{\tau}_n$, a local minimum of Equation 36 may not belong to the interval $[\tilde{\tau}_n \quad \tilde{\tau}_{n+1}[$; in this case, there could not be a synchronization peak at $d_f = \tilde{\tau}_n$: it is the phenomenon of synchronization peaks masking, as illustrated in Fig. 9 (b), with $\tilde{\tau}_0 = 0$, $\tilde{\tau}_1 = 0.0448$, $\tilde{\tau}_2 = 0.068$ and $\tilde{\tau}_3 = 0.3853$. By studying the behaviour of criterion F in the vicinity of points such that $d_f = \tilde{\tau}_n$, leads to the finding of the following condition:

$$\tilde{\tau}_{n-1} < \frac{n}{N_u^i} < \tilde{\tau}_n < \frac{n+1}{N_u^i} < \tilde{\tau}_{n+1}. \tag{37}$$

Equation 37 gives the necessary and sufficient condition of the synchronization peaks existence, as proved in (Nsiala Nzéza, 2006). Although the masking of peaks phenomenon occurs if the timing offsets of multiple users are close to each other in an asynchronous system, the synchronization can however be achieved combining the process described above with the modified-Pastd algorithm (Nsiala Nzéza, 2006). Thus, this consideration led us to analyze the MEV-based (MEVB) criterion described hereinafter.

5.1.2 MEVB criterion theoretical development

Without loss of generality, by setting $\sigma_b^2 = \beta = 1$ in Equation 24 leads to Equation 38, whose derivative with respect to the shift d_f is given by Equation 39:

$$\begin{cases} \lambda_n^0 = 2 - \langle d_f - \tilde{\tau}_n \rangle \\ \lambda_n^{-1} = 1 + \langle d_f - \tilde{\tau}_n \rangle \end{cases} \text{(a)}, \quad \begin{cases} \lambda_{n+1}^0 = 2 - \langle d_f - \tilde{\tau}_{n+1} \rangle \\ \lambda_{n+1}^{-1} = 1 + \langle d_f - \tilde{\tau}_{n+1} \rangle \end{cases} \text{(b)}, \tag{38}$$

$$\begin{cases} \dfrac{\partial \lambda_n^0}{\partial d_f} = -1 < 0 \\ \dfrac{\partial \lambda_n^{-1}}{\partial d_f} = 1 > 0 \end{cases} \text{(a)}, \quad \begin{cases} \dfrac{\partial \lambda_{n+1}^0}{\partial d_f} = -1 < 0 \\ \dfrac{\partial \lambda_{n+1}^{-1}}{\partial d_f} = 1 > 0 \end{cases} \text{(b)}. \tag{39}$$

Equation 39 (a) shows that for any shift $d_f \in [\tau_n, \quad \tau_{n+1}[$, λ_n^0 decreases while λ_n^{-1} increases, and conversely. The same result is obtained from Equation 39 (b), and the values for which they are equal are local minima. Therefore, the MEVB criterion is defined as the maximum value between two consecutive largest eigenvalues, which is equivalent to the following function $\mathcal{M}$:

$$\mathcal{M}(d_f) = \max_{\tau_n \le d_f < \tau_{n+1}} \left(\lambda_n^0, \lambda_{n+1}^{-1} \right), \; n = 0, \cdots, N_u^i - 1. \tag{40}$$

Since eigenvalues are linear functions of d_f, Equation 40 always presents maxima, and synchronization peaks at $d_f = \tilde{\tau}_n$, contrary to the FSNB criterion, as illustrated in Fig. 10, with the same parameters as those used in Fig. 9 (b). Thus, this discussion highlights that the performances obtained by using the MEVB criterion will be better than those obtained by using the FSNB criterion, as it will be shown in Section 5.3. Let us note that both criteria can easily be derived for downlink transmissions by setting $\tau_n = 0$, $n = 0, \cdots, N_u^i - 1$ in all equations above. At last, the number of synchronization peaks gives the number of interfering users. A particular emphasis has to be placed on the FSNB criterion due to some masking peaks. Moreover, once all peak positions are known, the differences between their positions and the corresponding user for which the normalized shift is the closest to 0 or 1, give an estimation of the transmission delays.

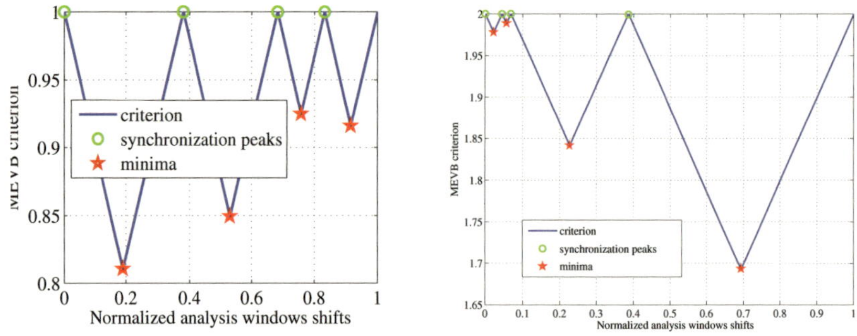

(a) ideal case (b) other cases

Fig. 10. Function $\mathcal{M}$ equivalent to the theoretical MEVB criterion in uplink, $N_u^i = 4$.

5.2 Sequences identification and symbols recovering

Once one of the interfering users within i is synchronised in uplink, e.g., $d_f = \tau_0$, which corresponds to set $\alpha_0 = 0$, the correlation matrix becomes:

$$\mathbf{R} = \sigma_b^2 \left\{ \beta \left(\mathbf{v}_{\alpha_0} \mathbf{v}_{\alpha_0}^* + \sum_{n=1}^{N_u^i - 1} \left\{ (1-\alpha_n) \mathbf{v}_n^0 (\mathbf{v}_n^0)^* \alpha_n \mathbf{v}_n^{-1} (\mathbf{v}_n^{-1})^* \right\} \right) + \mathbf{I} \right\}. \tag{41}$$

Equation 41 highlights a maximum eigenvalue which associated eigenvector contains the corresponding spreading sequence (excluding the effects of the global transmission filter), and $2(N_u^i - 1)$ eigenvalues. Then, this process is performed in an iterative way so as to get the N_u^i largest eigenvalues, which associated eigenvectors correspond to spreading sequences. In downlink, i.e., $\tau_n = 0$, $\alpha_n = \alpha$, $n = 0, 1, \cdots, N_u^i - 1$, the correlation matrix can be rewritten as:

$$\mathbf{R} = \sigma_b^2 \left\{ \beta \left(\sum_{n=0}^{N_u^i - 1} \left\{ (1-\alpha) \mathbf{v}_n^0 (\mathbf{v}_n^0)^* + \alpha \mathbf{v}_n^{-1} (\mathbf{v}_n^{-1})^* \right\} \right) + \mathbf{I} \right\}. \tag{42}$$

When users are synchronized, by using either the FSNB or the MEVB criterion, i.e., $\alpha = 0$, the induced corrlation matrix may be expressed as:

$$\mathbf{R} = \sigma_b^2 \left(\beta \sum_{n=0}^{N_u^i - 1} \mathbf{v}_n \mathbf{v}_n^* + \mathbf{I} \right). \tag{43}$$

Equation 43 exhibits the N_u^i largest eigenvalues whose associated eigenvectors correspond to the spreading sequences, and the $M - N_u^i$ others are on average equal to the noise power. Finally, linear algebra techniques previously described in Section 4, applied to the estimated eigenvectors, allow to identify sequences used at the transmitter and to recover transmitted symbols.

5.3 Simulation results in the multiuser multirate case

Simulations were carried out with complex GOLD sequences of processing gains L_0=31 and L_1=127. The other parameters were set as follows. The common chip frequency F_c=100 MHz, the sampling frequency F_e=300 MHz, T_F=2μs, K=300, N_u^0=2, N_u^1=4, and thus N_u=6.

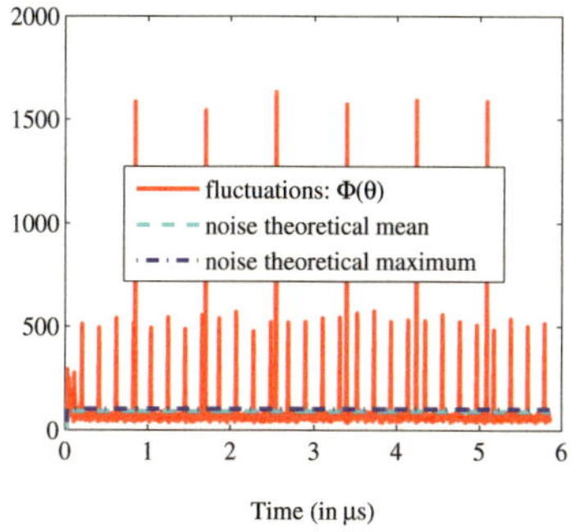

(a) Correlation fluctuations $\Phi(\tau)$

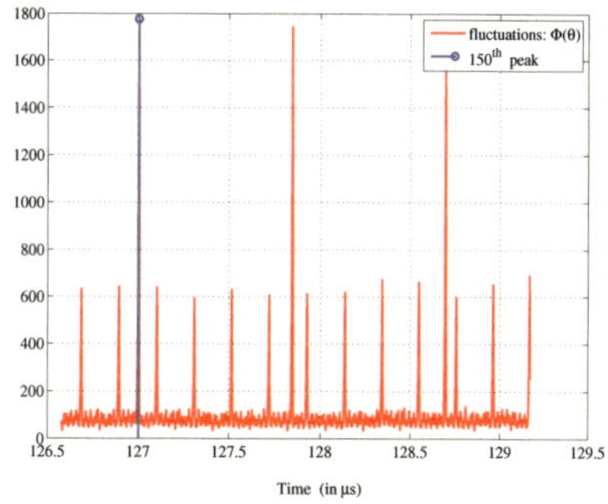

(b) Estimation of the symbol period T_{s_1}

Fig. 11. Estimation of the symbol period $N_u^0 = 2$, $N_u^1 = 4$, SNR= -7dB.

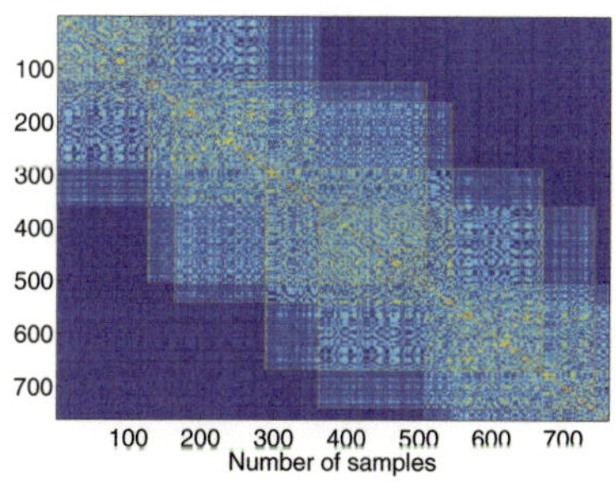

(a) Matrix $\mathcal{R}$ of double size before the sycnhronization

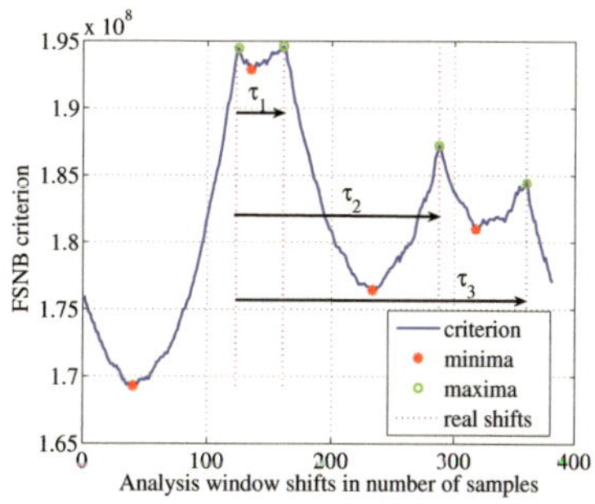

(b) Experimental FSNB criterion

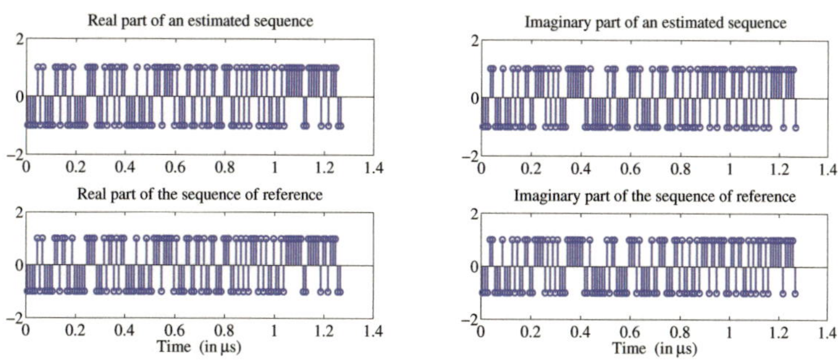

(c) Real and Imaginary parts of an estimated binary sequence

Fig. 12. Synchronization and sequences recovering, SNR= -7dB, $N_u^i = 4$, $L_1 = 127$.

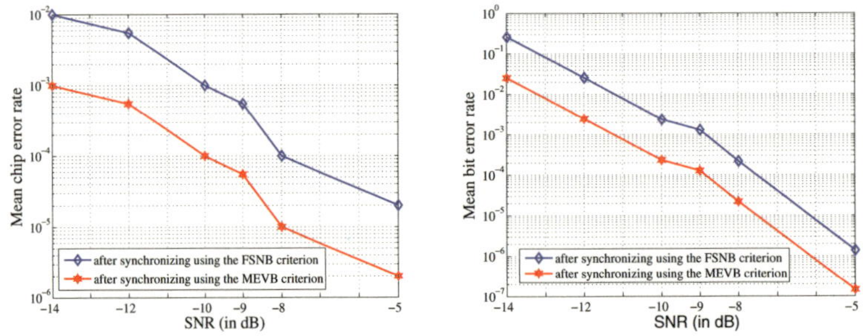

(a) Mean chip error rate (b) Mean bit error rate

Fig. 13. Both synchronization performances comparison for different SNRs, $N_u^1 = 4$, $L_1 = 127$.

A *QPSK* modulation was considered in uplink. For the N_u^1 users, it was set $\alpha_0 = 0.3228$, $\alpha_1 = 0.4226$, $\alpha_2 = 0.7533$ and $\alpha_3 = 0.9423$. Fig. 11 (a) illustrates fluctuations of the correlation estimator $\Phi(\theta)$. The curve clearly highlights two sets of equispaced peaks of different amplitudes. This means that two sets of spread spectrum signals transmitting at two different rates are hidden in the noise. Estimated symbol periods, (in μs) are $\widetilde{T}_{s_0} = 0.31$, $\widetilde{T}_{s_1} = 1.27$. To re-estimate these values, for each set of autocorrelation fluctuations, let us look for a maximum near the farthest peak on the right side, e.g., the 100th as shown in Fig. 11 (b), then we get $100 \times \widetilde{T}_{s_1} \approx 127 \mu s$.

Since symbol periods are available at this stage, e.g., $T_{s_1} = 1.27 \mu s$, the analysis window length is set as $T_F = T_{s_1}$. Fig. 12 (a) corresponding to the double size matrix $\mathcal{R}$ shows 4 overlapping matrices, suggesting that there are 4 interfering users. Matrix $\mathbf{R}$ before the synchronization corresponds to the $(M \times M)$-matrix extracted from $\mathcal{R}$. Fig. 12 (b) shows the experimental FSNB criterion calculated by moving along the diagonal of matrix $\mathcal{R}$ and calculating the squared norm of the shift-induced submatrix for each value of the shift. This squared norm will be maximum for the bright areas in Fig. 12 (a).

Fig. 12 (b) evidences 4 peaks which X-coordinates give the desynchronization times in number of samples. The number of peaks gives the number of active users over the analysis window, hence there are 4 interfering active users. By taking the oversampling factor into account, we obtain $\widehat{\alpha}_0 \approx 0.3281$, $\widehat{\alpha}_1 = 0.4226$, $\widehat{\alpha}_2 = 0.7533$ and $\widehat{\alpha}_3 = 0.9449$.

Since transmission delays and shifts are supposed to be sorted in the ascending order, the first peak, i.e., for $\widehat{\alpha}_0$, corresponds to the reference user. Thus, $\widehat{\tau}_1 = 0.0999$, $\widehat{\tau}_2 = 0.4252$, and $\widehat{\tau}_3 = 0.6221$. These values are very close to the real ones, and allow the synchronization of interfering users. Fig. 12 (c) shows the real and imaginary parts of one of the estimated sequences, which are equal to that used at the transmitter side, respectively.

Fig. 13 (a) highlights the performances in terms of mean chip error rate (MCER), i.e., the average ratio of the number of wrong sequence chips to the total number of sequence chips. It shows very low MCERs after synchronizing using either the MEVB or FSNB criterion. Moreover, MECRs obtained after using the MEVB criterion are lower than those obtained after using the FSNB one (with a 3 dB gain), in agreement with the theoretical analysis. It also shows that in average, one chip at most is wrong with sequences of length 127. Fig. 13 (b) shows the performances in terms of mean bit error rate (MBER). It clearly evidences very good

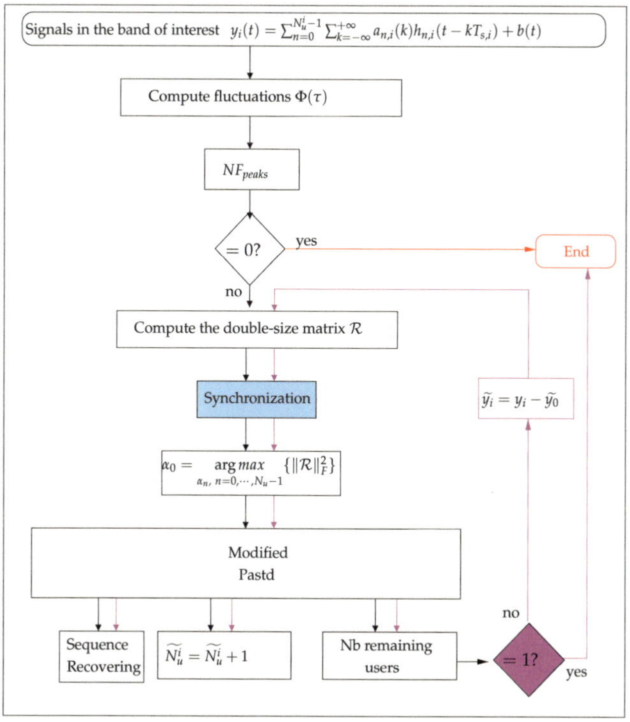

Fig. 14. Iterative implementation of the proposed scheme in the multiuser single rate case.

performances after synchronizing using both criteria and estimating transmitted symbols. In agreement with the results shown in Fig. 13 (a), lower MBERs are obtained (with a 2 dB gain) after using the MEVB criterion. Typically, in some cooperative systems, the MBER is about 10^{-8} which is very close to that obtained at -5 dB after using the MEVB criterion.

6. Performances improvement of the proposed method

Note first that the overall chain is implemented in a way to make it interactive with an operator. As claimed in Section 3, although the proposed method has been described sequentially, the most efficient implementation of the three steps is to link them back iteratively. This allows the global chain performances improvement by deflating the estimated signal. This will be especially important when the number of interfering users within a given group of users transmitting at the same data rate increases, or when the differences between data rate from one group to another one is very high. In these cases, some synchronization peaks can be masked as proven in Section 5. Hence, it is advisable to adopt the iterative approach summarized on Fig. 14 and Fig. 15. Therefore, firstly identifying in the multiuser detection step the group of users corresponding to the largest sequence or fluctuations amplitude peaks, set to NF_{peaks}.

6.1 Implementation in the multiuser single rate case

This case is shown on Fig. 14. Within the identified group of users among NF_{peaks}, the user corresponding to the synchronization criterion maxima is first synchronized. Besides, the

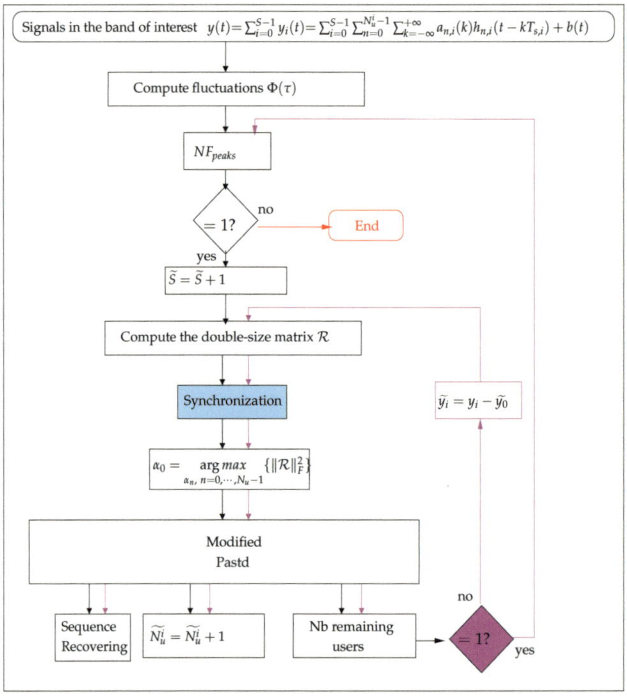

Fig. 15. Iterative implementation of the proposed scheme in multiuser multirate case

number of synchronization peaks set to NS_{peaks} must be stored for further comparison in the algorithm. In the next step, the corresponding spreading sequences estimated will be the most reliable, since it will correspond to the eigenvector, thus the signal space associated with the largest eigenvalue.ĂľAt this stage, using the modified-Pastd algorithm aiming to estimate the number of users $\widetilde{N}_u^i$ is necessary.ĂľThe identified signal is substracted from the global signal corresponding to signals transmitted at the same data rate. Then, if the number of remaining users Nb is not equal to one, the synchronization criterion may be computed again after the deflation, and the process is repeated as described on Fig. 14.

6.2 Implementation in the multiuser multirate case

The iterative implementation in the multiuser multirate context is illustrated on Fig. 15. In this case, first identify the group of users corresponding to the largest fluctuations amplitude peaks, and store the number NF_{peaks} of group of fluctuations peaks. Then, the number of iteration compared to NF_{peaks} is increased. Therefore, if $Nb \neq 1$, restart the process by selecting another group of fluctuations. Note that, improving the performances is achueved by recomputing the function $\Phi(\theta)$ after deflating the estimates of the signals whithin the identified group of interest.

6.3 Discussion and illustration through simulation results

In asynchronous multiuser multirate CDMA systems, the near-far effect is definitely present. Consequently, the multipath channels and/or under near-far effects may be mitigated through the iterative implementation of the proposed algorithms highlighted above. Indeed

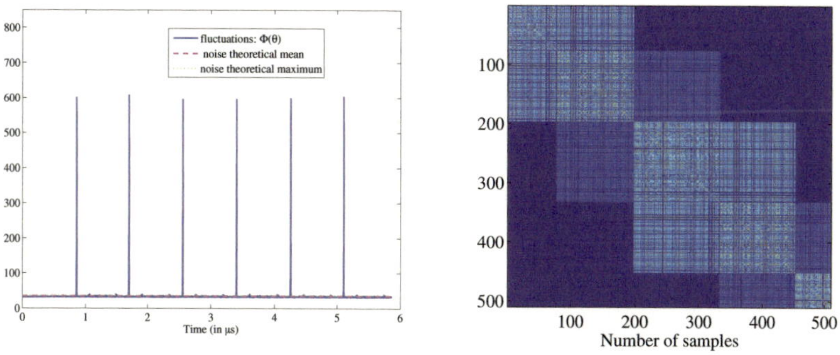

(a) Fluctuations $\Phi(\tau)$ (b) Double-size matrix $\mathcal{R}$

Fig. 16. Fluctuations and double-size correlation matrix analysis.

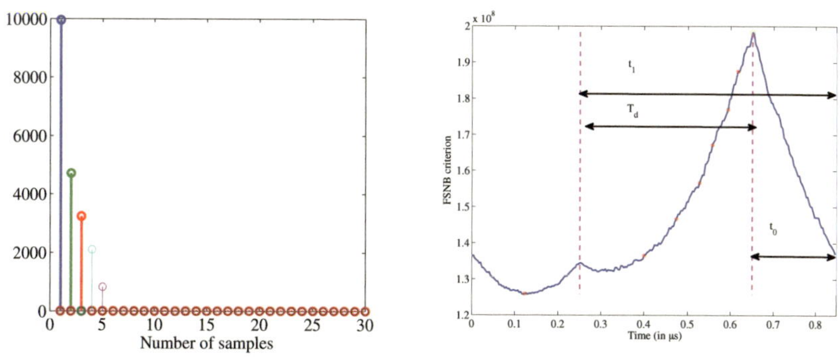

(a) Eigenvalues before the synchronization (b) Synchronization criterion

Fig. 17. Illustration of the synchronization process.

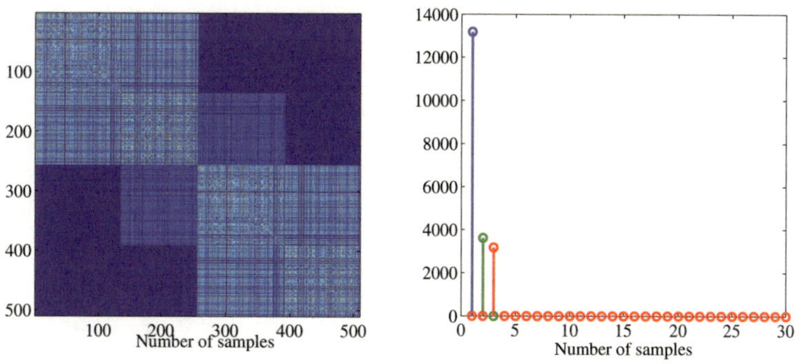

(a) Extracted matrix $\mathbf{R}_{\widetilde{\alpha}_0}$ (b) Eigenvalues after the synchronization

Fig. 18. Illustration of the third step of the proposed scheme.

this iterative approach jointly uses the modifed-Pastd algorithm and a feedback on either synchronization criteria or fluctuations of correlation estimators computation. Besides, note that the deflation technique is widely used in CDMA signals detection in both blind and cooperative methods. The originality of our approach lies in the analysis of the fluctuations of correlation estimators, and in the synchronization criteria implemented.

As a result, the proposed algorithms may be applied to the long sequence code case, as investigated in Section 7 through simulation results.

Let us now illustrate the iterative implementation using the FSNB criterion with following simulations parameters were set: $N_u^0 = 2$ complex GOLD sequences of processing gain $L_{0,0} = L_{1,0} = 255$, with inequal powers: $\sigma_{s_0}^2 = 1$ and $\sigma_{s_1}^2 = 0.5022$, $\sigma_b^2 = 5.3307$, $F_c = 300$ MHz, $F_e = 600$ MHz, $T_F = 2\mu s$, $K = 894$, a $QPSK$ modulation was considered in uplink. For the N_u^0 users, it was set $t_0 = 0.2$ μs, $t_1 = 0.6$ μs, $\text{SNR}_0 = -6.3$ dB, $\text{SNR}_1 = -9.78$ dB, what leads to a global SNR$\simeq -11$ dB at the receiver side. From Fig. 16 (a), we get $\widetilde{T_S} = 0.85$ μs, which allows to compute the double-size matrix $\mathcal{R}$, as shown on Fig. 16 (b). Even though on Fig. 16 (a), an estimation was not possible for N_u^0, the analysis of Fig. 16 (b) suggests that there could be $\widetilde{N}_u^0 = 2$ two users.

This is also reinforced by the eigenvalue decomposition that reveals 4 largest eigenvalues, as illustrated on Fig. 17 (a). However, the synchronization criterion on Fig. 17 (b) exhibits only one peak, from which we get $\widetilde{t}_0 = 0.208$, and the estimates $\widetilde{T_d} = 0.342$ μs. The EVD on Fig. 17 (b) of the corresponding extrated matrix $\mathbf{R}_{\widetilde{\alpha}_0}$ on Fig. 18 (a) exhibits 3 largest eigenvalues as expected. Since the use of the modified-Pastd algorithm leads to $\widetilde{N}_u^0 = N_u^0$, the contibution of the estimated signal is substracted from the global received signal and we continue by recomputing the criterion in order to estimate the second user.

Computing again another double-size matrix as shown on Fig. 19 (a). Its EVD on Fig. 19 (b) leads to only two largest eigenvalues as expected. Besides, the obtained criterion in this second iteration on Fig. 20 (a) exhibits only one synchronization peak, which allows to synchronize the corresponding user. Furthermore, the sequence recovering process may be performed, starting with the eigenvector corresponding to the remaining largest eigenvalue as shown on Fig. 20 (b).

7. Application to long spreading sequences transmission case

In order to clear up the reader, the application of the proposed method will be done through simulations in uplink. The multiuser case can be easily deduced as it will be further shown. Following parameters were set: Complex GOLD sequence of Length L_0=63, Q_s=3 which

implies $Q_s T_s = L T_c$ (see Subsection 2.1.2 for details), $BPSK$ modulation α_0=0.2, F_e=300 MHz, F_c=150 MHz, T_F=2 μs,K=666, SNR $= -3$ dB. Fig. 21(a) illustrates double-size matrix $\mathcal{R}$ computed with the estimates $Q_s * \widetilde{T_S}$=1.20 μs in this case. The fluctuations curve is not represented here since it the same as in the short code case. The great difference is in estimating $Q_s * \widetilde{T_S}$=1.2588 μs rather than $\widetilde{T_S} = L/F_c$=0.42 μs, and we get $\widetilde{\alpha}_0$=0.1667 which is very close to the real value. Seeing $2 * Q_s$=6 overlapping subspaces as expected. Compared to the short code case,i.e., Q_s=1, obtaining $2 * Q_s$=2 overlapping subspaces, as detailed in Section 4.

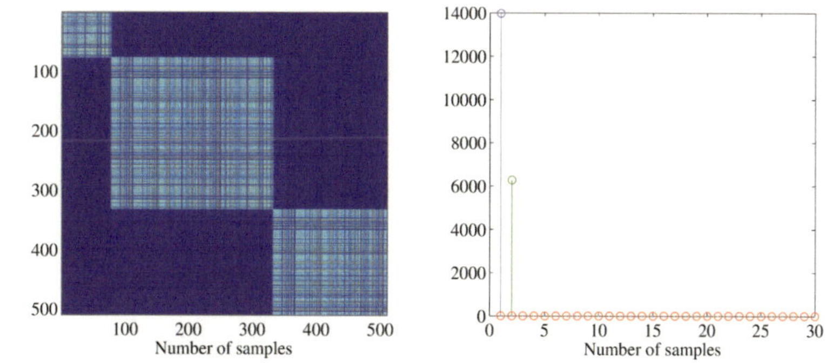

(a) Double-size matrix $\mathcal{R}$

(b) Eigenvalues before the synchronization

Fig. 19. Subspace analysis before the synchronization process in the second loop after deflation.

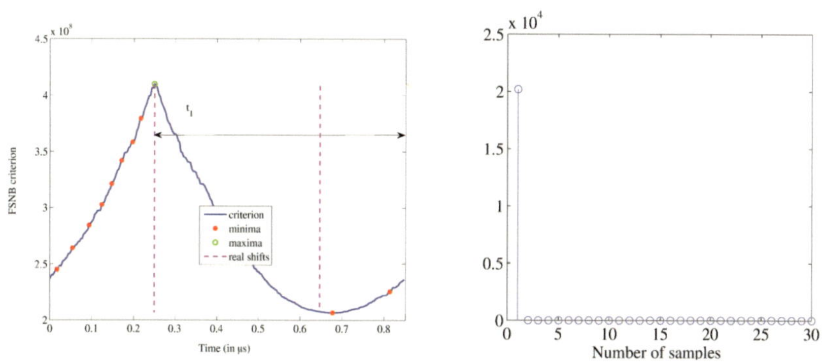

(a) Synchronization criterion

(b) Eigenvalues after the synchronization

Fig. 20. Synchronization process in the second loop after deflation.

Besides, the criterion synchronization on Fig. 21 (b) proves that there are Q_s=3 subspaces corresponding to the Q_s=3 part of the spreading sequence. Therefore, synchronizing starting from the criteria maximum allows the extraction of the corresponding sub-matrix as indicated on Fig. 21 (c), on which linear algebra techniques application described in Section 4 lead to the estimation of a part the spreading sequence. The process is done Q_s-1 times in order to estimate the whole spreading code as depicted on Fig. 21 (d).

To complete, in the multiuser context, theoretical developments in Section 5 remain valid for the long case. Indeed, let us have a look on Fig. 22 obtained with the same parameters than on Figures 16 and 17, but with Q_s=5 for the two users (N_u^0=2). As evidenced on Fig. 22 (a) $N_u^0{*}2{*}Q_s$=20 overlapping subspaces, which suggests that there are two interfering users with long spreading codes. However, $(N_u^0-1){*}2{*}Q_s$=10 overlapping subspaces are not very remarkable as they correspond to the user received with a very low power. This is confirmed by the synchronization criterion plotted on Fig. 22 (b) which only emphasizes the shifts of the signal received with the highest power, i.e., the equispaced maxima of all the synchronization maxima. Fig. 22 (b) highlights the synchronization peaks masking. Hence,

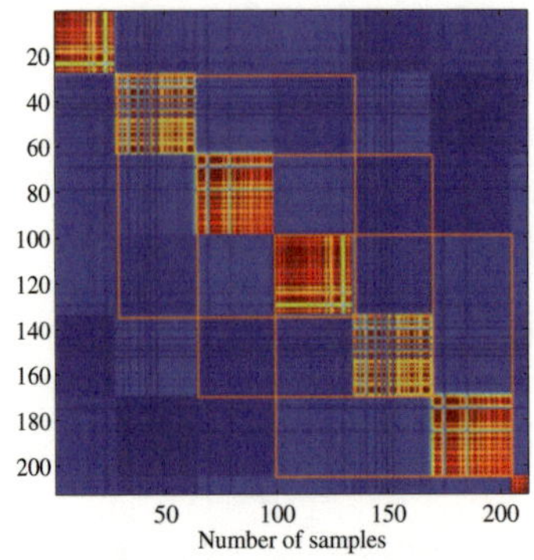

(a) Double-size matrix $\mathcal{R}$

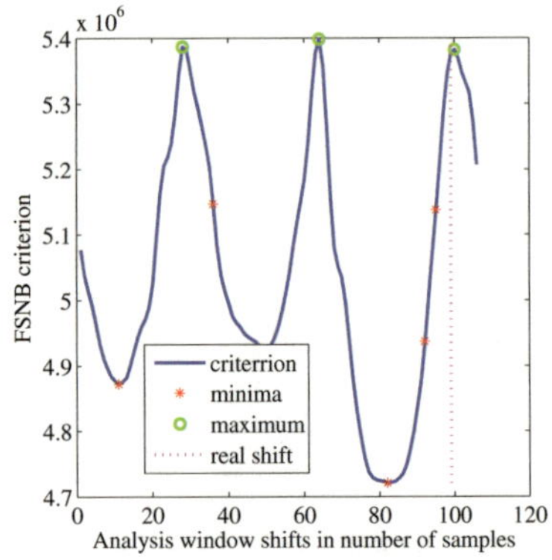

(b) Synchronization criterion

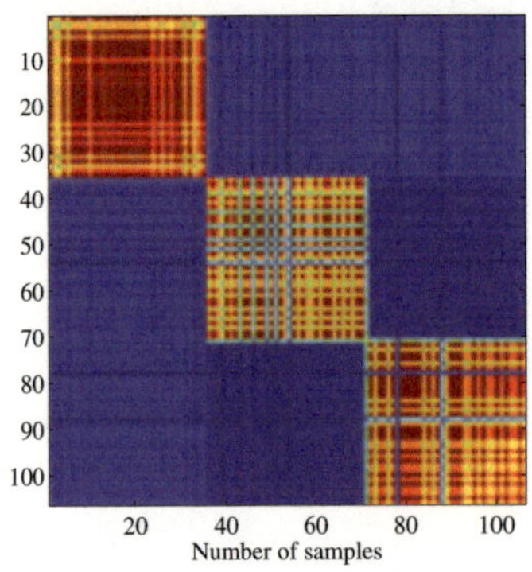

(c) Extracted matrix $\mathbf{R}$

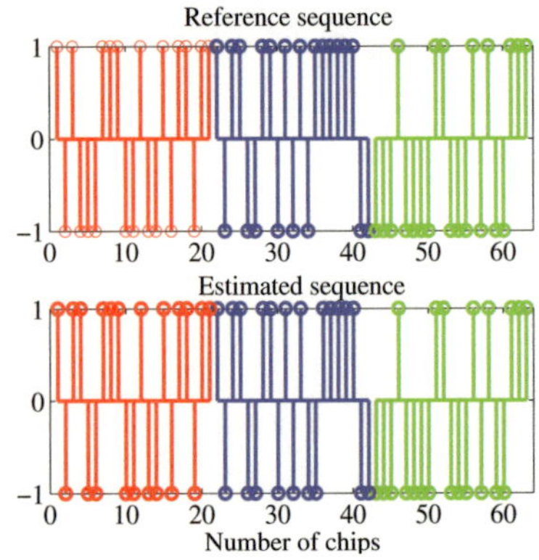

(d) Sequence estimation after *binarization*

Fig. 21. Application of the proposed algorithm to the single user long code case in uplink.

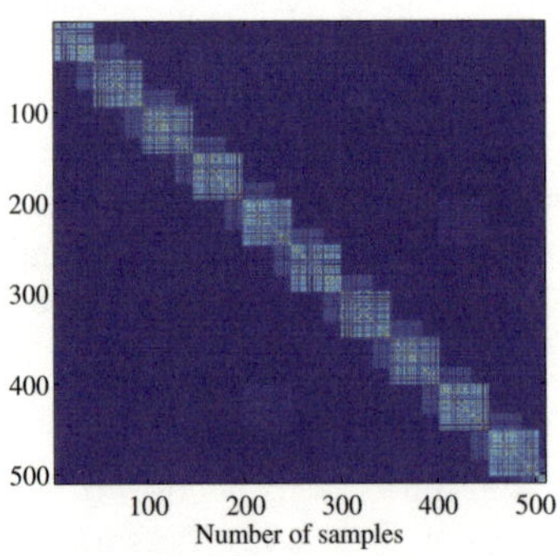

(a) Double-size matrix$\mathcal{R}$

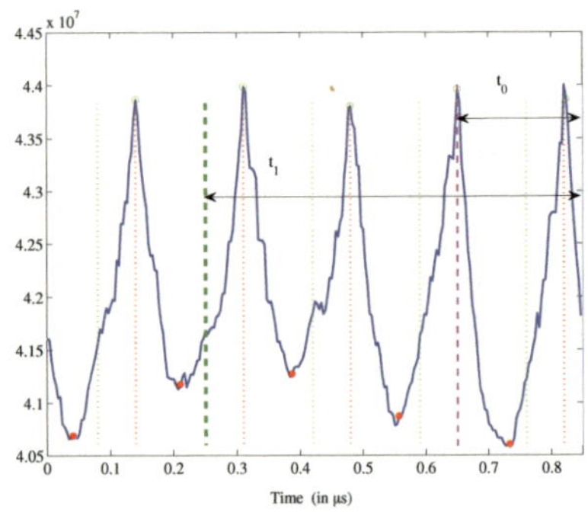

(b) Synchronization criterion

Fig. 22. Illustration of the synchronization process with $N_u^0 = 2$ users in uplink.

the process described above for the case on Fig. 21 may be performed in order to recover the corresponding spreading sequences for this user. Furthermore, after the deflation of the user as we just estimated, and recomputing the synchronization criterion in order to estimate the second interfering user as depicted in Subsection 6.3.

8. Chapter summary and conclusion

In this chapter, we described a fast and efficient blind spread spectrum signal analysis approach consisting on signal detection, synchronization and identification of all concomitant data rates and user's sharing the same bandwidth in DS-CDMA transmissions. All this signal processing bricks are dedicated to the self-reconfiguration of Cognitive Radio terminals. We have evidenced that this approach permits to differentiate various CDMA standards through the assessment of interfering signals data rate. The methods and algorithms we have developed for these three steps allow to determine CDMA signals parameters: number of standards or rates, number of users for each rate, type of long or short sequences, the resynchronization time, spreading sequences and to despread signals of each user. We have shown that even in a sequential implementation,the approach developed and implemented here is very efficient both in terms of detection, synchronization and identification. Moreover, its performances improvement has been addressed thourgh two synchronization criteria. The first one has a very low computational cost and the second one permit to greatly reduce the phenomenon of peak masking, related to resynchronization times too close or excessive power difference between two users. Alternatively, a proposed iterative implementation that can suppress almost all the problems of peak masking by successive subtractions of interfering users was detailed. From which, short or long codes DS-CDMA signals blind detection and identification may be performed. Finally, the major part of the signal processing block of this approach were implemented on an operational system for the monitoring of radio frequency spectrum.

9. References

Buzzi, S., Venturino, L., Zappone, A. & Maio, A. D. (2010). Blind user detection in doubly dispersive ds/cdma fading channels, *IEEE Transaction on Signal Processing* 58(3): 1446–1451.

Cheraghi, P., Ma, Y. & Tafazolli, R. (2010). A novel blind spectrum sensing approach for cognitive radios, *PGNET 2010 Conference*, Marrakech, Morroco.

Committee, C. . M. (1989). *Digital land mobile radio communications : final report*, Euratom publications - EUR 12160 EN, Information technologies and sciences, *Commission of the European Communities*.

Ghavami, S. & Abolhassani, B. (2008). Blind chip rate estimation in multirate cdma transmissions using multirate sampling at slow flat fading channels, *IEEE ICCS 2008*, Singapore, pp. 1344– 1348.

Ghavami, S. & Abolhassani, B. (2009). On the performance of blind chip rate estimation in multi-rate cdma transmissions using multi-rate sampling in slow flat fading channels, *Wireless Sensor Network* 2: 67–75.

Hosseini, S., H, A. & J.A., R. (2010). A new cyclostationary spectrum sensing approach in cognitive radio, *IEEE SPAWC*, Marrakech, Morroco, pp. 1–4.

Khodadad, F. S., Ganji, F. & Mohammad, A. (2010). A practical approach for coherent signal surveillance and blind parameter assessment in asynchoronous ds-cdma systems in multipath channel, *Iranian Conference on Electrical Engineering*, Isfahan, Iran, pp. 305–310.

Khodadad, F. S., Ganji, F. & Safaei, A. (2010). A robust pn length estimation in down link low- snr ds-cdma multipath channels, *IEEE International Conference on Advanced Communication Technology* , Angwon-Do, South Korea, pp. 951–955.

Kibangou, A. Y. & de Almeida, A. (2010). Distribured parafac based ds-cdma blind receiver for wireless sensor networks, *IEEE SPAWC*, Marrakech, Morroco, pp. 1–5.

Koivisto, T. & Koivunen, V. (2007). Blind despreading of short-code ds-cdma signals in asynchronous multi-user systems, *Signal Processing (87)* 87(11): 2560–2568.

Meng, Y., You, M.-L., Luo, H.-W., Liu, G. & Yang, T. (2010). The linearly constrained lscma for blind multi-user detection, *Wireless Personal Communications: An International Journal* 53(2): 199–209.

Mitola, J. (2000). *Cognitive Radio: An Integrated Agent Architecture for Software Defined Radio*, PhD thesis, Royal Institute of Technology (KTH), Stockholm, Sweden.

Nsiala Nzéza, C. (2006). *Récepteur Adaptatif Multi-Standards pour les Signaux à Étalement de Spectre en Contexte Non Coopératif*, PhD thesis, Université de Bretagne Occidentale, Brest, France.

Nsiala Nzéza, C., Gautier, R. & Burel, G. (2004). Blind synchronization and sequences identification in cdma transmissions, *IEEE-AFCEA-MilCom*, Monterey, California, USA.

Nzéza, C. N., Gautier, R. & Burel, G. (2006). Blind Multiuser Detection in Multirate CDMA Transmissions Using Fluctuations of Correlation Estimators, *IEEE Globecom 2006*, San Francisco, California, USA, pp. 1–5.

Nzéza, C. N., Gautier, R. & Burel, G. (2008). Theoretical Performances Analysis of the Blind Multiuser Detection based on Fluctuations of Correlation Estimators in Multirate CDMA Transmissions, *Military Technical Academy Journal*.

Nzéza, C. N., Moniak, G., Berbineau, M., Gautier, R. & Burel, G. (2009). Blind MC-DS-CDMA parameters estimation in frequency selective channels, *IEEE-GlobeCom-5th IEEE Broadband Wireless Access Workshop*, Honolulu, Hawaii, USA.

Shen, L., Wang, H. & andZhijin Zhao, W. Z. (2011). Blind spectrum sensing for cognitive radio channels with noise uncertainty, *IEEE Transaction on Wireless Communications* 10(6): 1721–1724.

Williams, C., Beach, M., Neirynck, D., A. Nix, K. C., Morris, K., Kitchener, D., Presser, M., Li, Y. & McLaughlin, S. (2004). Personal area technologies for internetworked services, *IEEE Communication Magazine* 42(12): 15–32.

Yu, M., Chen, J., Shen, L. & Li, S. (2011). Blind separation of ds-cdma signals with ica method, *Journal of Networks* 6(2): 198–205.

Zhang, T., Dai, S., Zhang, W., Ma, G. & Gao, X. (2011). Blind estimation of the pn sequence in lower snr ds-ss signals with residual carrier, *Digital Signal Processing In Press* .

7

Measurement and Statistics of Spectrum Occupancy

Zhe Wang
Durham University
UK

1. Introduction

Based on the conception of spectrum sharing, Cognitive Radio as a promising technology for optimizing utilization of the radio spectrum has emerged to revolutionize the next generation wireless communications industry Staple & Werbach (2004) Ashley (2006). In order to facilitate this technology, the present spectrum allocation strategies have to be re-examined and the actual spectrum occupancy information has to be studied systemically. To assess the feasibility of Cognitive Radio technology, the statistical information of the current spectral occupancy needs to be investigated thoroughly.

While preliminary occupancy information can be retrieved from spectrum licences, essential details are often unknown generally, which include the location of transmitters,transmitter output power, and antenna type, etc. Additionally, licences do not specify how often the spectrum is being occupied. Furthermore, the local environment affects the propagation of radio waves a lot. While these affects can be simulated, the results are hardly precision. Hence, in order to categorize spectrum usage, practical spectrum monitoring are vastly preferable to theoretical analysis Carr (1999).

Two important characteristics of the spectrum are the propagation features and the amount of information which signals can carry. In general, signals sent using the higher frequencies have smaller propagation distances but a higher data carrying capacity. These propagation characteristics of the spectrum constrain the identified range of applications for which any particular band is suitable Saakian (2011). A portion of spectrum range from 30-3000 MHz is known to be suitable for a wide variety of services and is thus in great demand, which became the main investigation in our project.

We studied the 100-2500 MHz spectrum with the radio monitoring systems which technical details have been fully recorded in this article. In this chapter,we will present the detail statistics of spectrum occupancies with graphics and tables, which give the overall profile of current spectrum usage in this spectrum. The conclusion of the statistical information from the spectrum monitoring experiments shows that the spectrum occupancy range from 100-2500 MHz are low indeed in the measuring locations and period. The average occupancies for most bands are less than 20%. Especially, the average occupancies in the 100-2500 MHz spectrum are less than 5%. This suggests that Cognitive Radio technology can physically play an important role in future communications, if the current radio spectrum allocation strategies would be modified Wang & Salous (2009).

2. Radio spectrum occupancy monitoring

2.1 Considerations for spectrum monitoring

Cognitive radio technologies require reconsideration the current regulations and policies of spectrum management. Spectrum monitoring is a fundamental function to support spectrum management. Spectrum measurements are critical to policy regulators and researchers in the development of new spectrum access technologies. Specifically, spectrum occupancy studies inspect what spectrum bands have low or no active utilization and thus may be appropriate for spectrum sharing. They also provide information on the signal characteristics within these bands, which is needed to design spectrum sharing algorithms. Referred to ITU (n.d.) the considerations of spectrum monitoring activities in our project include the following:

- that licensed user information from the frequency management databases only indicates that the use of the frequency is authorized. The number of assignments on a frequency does not give any actual use information of that particular frequency.
- that efficient Spectrum Management can only satisfactorily proceed if the monitoring information provides the radio spectrum regulators with adequate reliable information about the actual usage of the spectrum.
- that results of spectrum occupancy measurements will give information about the current use of frequencies to establish that the spectrum is being used efficiently and to assess the feasibility of the new technologies.

The overall goal of spectrum monitoring activities of our project is to depict the current levels of spectrum usage in the range 100 to 2500 MHz and its implications for cognitive radio. Central objectives include the following:

- to provide information of spectrum efficiency for determining planned and actual frequency usage and occupancy, and for assessing the feasibility of spectrum sharing technique.
- to provide data for statistical modeling.

The measuring system should be chosen carefully to ensure capabilities exist with the spectrum management agency to effectively monitor and analyze the frequency bands.

2.2 Monitoring system and site

A successful spectrum survey requires careful selection of a measurement site. The location chosen for measuring will affect measured spectrum occupancy. For example, measurements made in Durham are probably representative of many towns that have similar scale and do not have heavy military activity or maritime radio navigation etc. Generally, a site for spectrum monitoring requires Sanders et al. (1996):

- "limited numbers of nearby transmitters to prevent intermodulation or saturation problems that can arise even though pre-selection and/or filtering is used for survey measurements
- limited man-made noise such as impulsive noise from automobile ignition systems and electrical machinery that can add to the received signals of interest."

Figure 1 shows that the measurement locations for the spectrum occupancy project was the roof and inside of Engineering Building, Durham University.

Fig. 1. Measurement location

Because of the complexity and sophistication of wireless communication technologies, it is an ever-increasing challenge to monitor the spectrum, particularly considering the rapid growth of wireless, satellite, and point-to-point communication devices. Key considerations in the design of spectrum monitoring systems include types of equipment, data rate and complexity of data capture and processing, degree of integration with software tools for analysis. While considered the limited project budget and the existing equipment, we integrated independently 3 different spectrum monitoring systems for different frequency bands to satisfy the technique requirements.

The monitoring system for 100-1500 MHz spectrum in this project was configured as Figure 2. The system consisted of an omnidirectional Dressler ARA-1500 active antenna range from 50 to 1500 MHz connected by a 6 m RF cable to a diplexer RSM-2000 which allows the DC current for the preamplifier to be applied to the centre conductor of the RF coax, eliminating the need for an additional DC power feed conductor. The RSM-2000 also contains a 20 dB RF adjustable attenuator allowing received signals to be attenuated over the entire frequency range. In order to increase the dynamic range of the system, a highpass filter Mini-Circuits SHP-100 was connected to the output of the RSM-2000. A Jim M-75 low noise amplifier was inserted in front of HP 8560A spectrum analyzer to decrease its noise figure. The GPIB bus was used for logging the trace data onto the hard disk of a PC and for transferring control command sequences to the spectrum analyzer. Table 1 lists the configuration parameters of spectrum analyzer for scanning 100-1500 MHz. Justifications will be given in the next section.

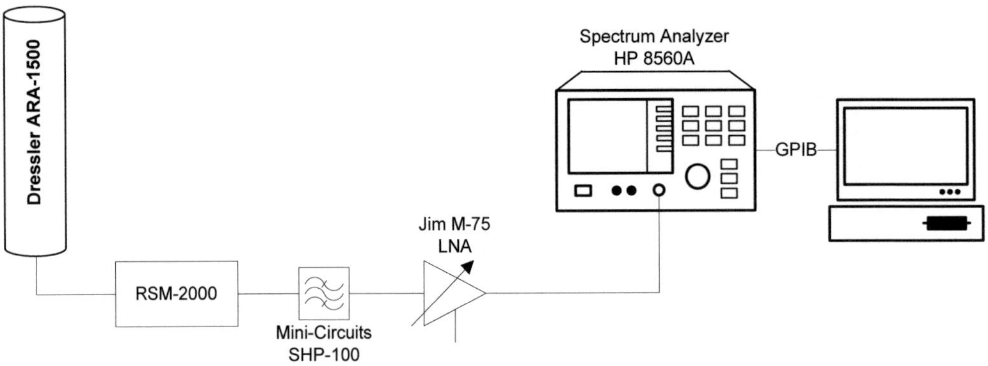

Fig. 2. Monitoring system for 100-1500 MHz spectrum

Model	HP 8560A
Frequency span (MHz)/ sweep	6
Resolution bandwidth (kHz)	10
Sweep time (s)	6
Detection mode	Sample
RF attenuator (dB)	10
Reference level (dBm)	-20

Table 1. Configuration parameters for scanning 100-1500 MHz

The monitoring system for 1500-2500 MHz spectrum was configured as in Figure 3. Instead of an omnidirectional antenna, an antenna array consisted of 4 directional Log-periodic antennas which enable the detection of signal incident directions. An RF switch controlled by the PC parallel port was used to choose a given antenna. Table 2 lists the configuration parameters of spectrum analyzer for scanning 1500-2500 MHz. Justifications will be given in the next section.

Model	HP 8560A
Frequency span (MHz)/ sweep	12
Resolution bandwidth (kHz)	10
Sweep time (s)	6
Detection mode	Sample
RF attenuator (dB)	10
Reference level (dBm)	-20

Table 2. Configuration parameters for scanning 1500-2500 MHz

3. Statistics of spectrum occupancy

It is often helpful to make some simple characterization of the data in terms of summary statistics and graphics. The spectrum measurements contained in this chapter can only be used to assess the feasibility of using alternate services or systems under restricted conditions. Extrapolation of data in this paper to general spectrum occupancy for spectrum sharing requires consideration of additional factors. These include spectrum management regulations, types of missions performed in the bands and new spectrum requirements in the development and procurement stages. Also, measurement area, measurement site, and measurement system parameters should be considered Freedman et al. (2007).

Highly dynamic bands where occupancy changes rapidly include those used by mobile radios (land, marine, and airborne) and airborne radars. These bands should be assigned a high priority and be measured often during a spectrum survey in order to maximize opportunities for signal detection. Bands that are not very dynamic in their occupancy such as those occupied by commercial radio and television signals or fixed emitters such as air traffic control radars need not be observed as often, because the same basic occupancy profile will be generated every time. Such bands should be given a low priority and less measurement time. An extreme case is that of the common carrier bands, which are essentially non-dynamic.

Boxplot in Figure 4 also known as a box-and-whisker diagram Freedman et al. (2007) is a convenient way of graphically depicting groups of numerical data. A boxplot shows a measure of central location (the median), two measures of dispersion (Q_1 and Q_3[1] and inter-quantile range IQR), the skewness (from the orientation of the median relative to the quartiles) and potential outliers (marked individually). Boxplots are especially useful when comparing two or more sets of data. Figure 5 shows overall occupancy statistics of each band in the frequency range 100 MHz to 2500 MHz, where the threshold was set to -100 dBm. Figures 6 to 16 describe spectrum occupancy measurements and statistics of each band in the frequency range 100 MHz to 2500 MHz in Durham area during the period of 27/06/07 – 03/07/07. The spectrum occupancy in the *frequency domain* is shown in the top panel. This panel shows *Average* with time, in which the power values of each 10 kHz channel are linearly averaged during the measurement period, and *Maximum*, in which the result for any given

[1] The quantile function is the inverse of the cumulative distribution function. The p-quantile is the value with the property that there is probability p of getting a value less than or equal to it. Q_1 is 25-quantile, Q_3 is 75-quantile.

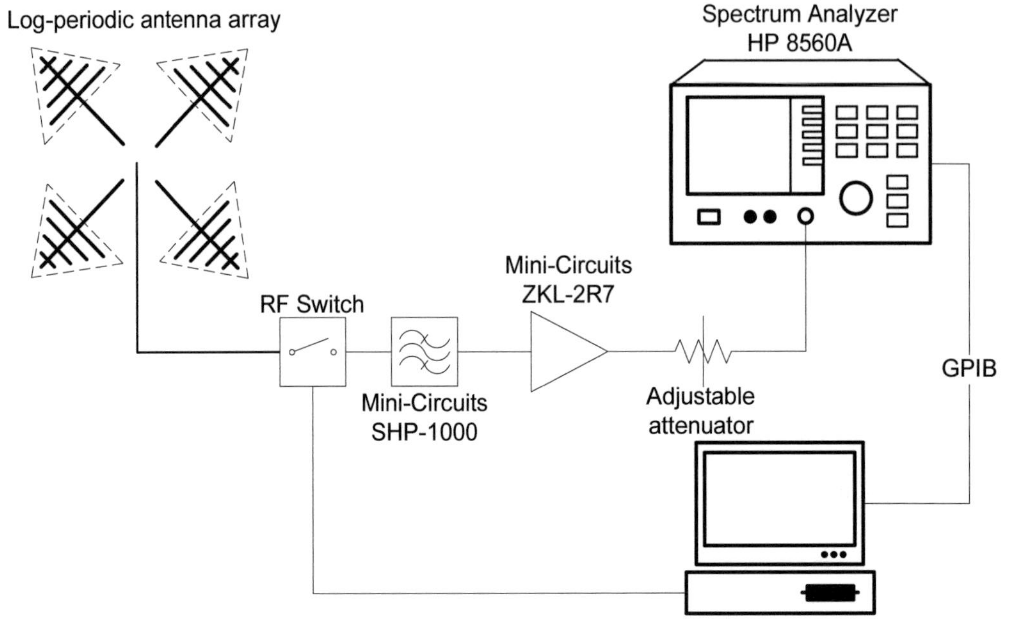

Fig. 3. Monitoring system for 1500-2500 MHz spectrum

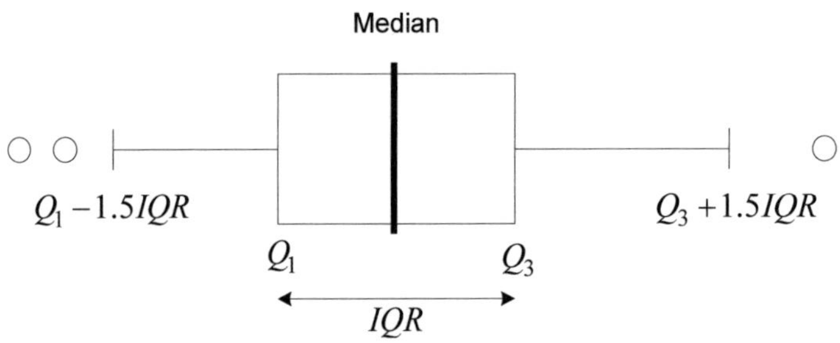

Fig. 4. Box-and-whisker diagram

channel is the maximum power ever observed in that channel in 7-day time. Together, the *Average* and *Maximum* results provide a simple characterization of the temporal behavior of a channel. For example, when the results are equal, it suggests a single transmitter which is always on and which experiences no fading. At the other extreme, a large difference between the mean and maximum measurements suggests intermittent use of the channel.

The middle panel shows the band occupancy in the *time domain* with thresholds -95 dBm and -100 dBm during the measurement period of 27/06/07 – 03/07/07. For example, the occupancy rate of the Air band shown in Figure 6 was calculated in each time point in a given threshold -95 dBm and -100 dBm respectively. Total 168 time points for each hour of 7-day were plotted in this panel.

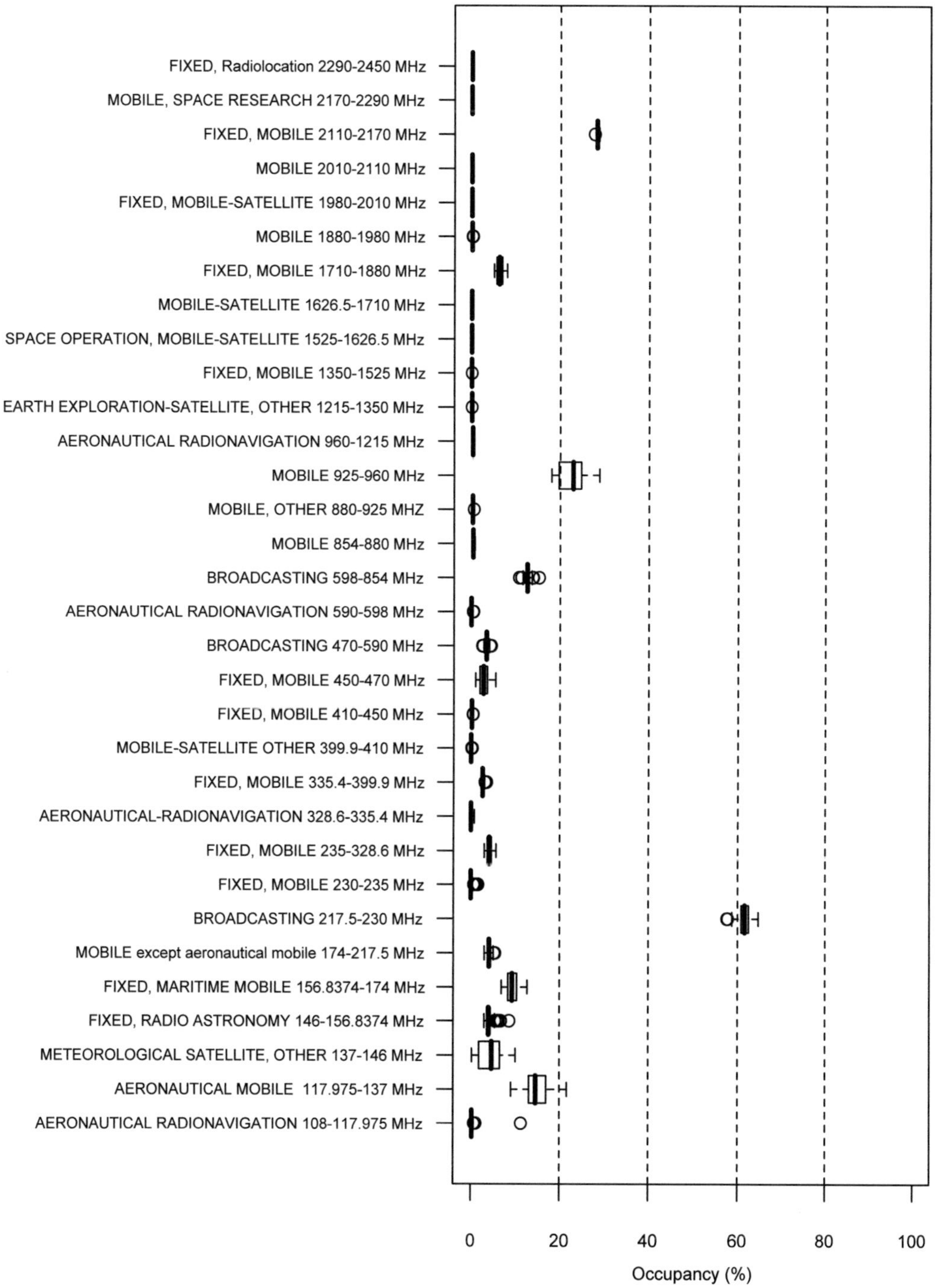

Fig. 5. Occupancy statistics of spectrum

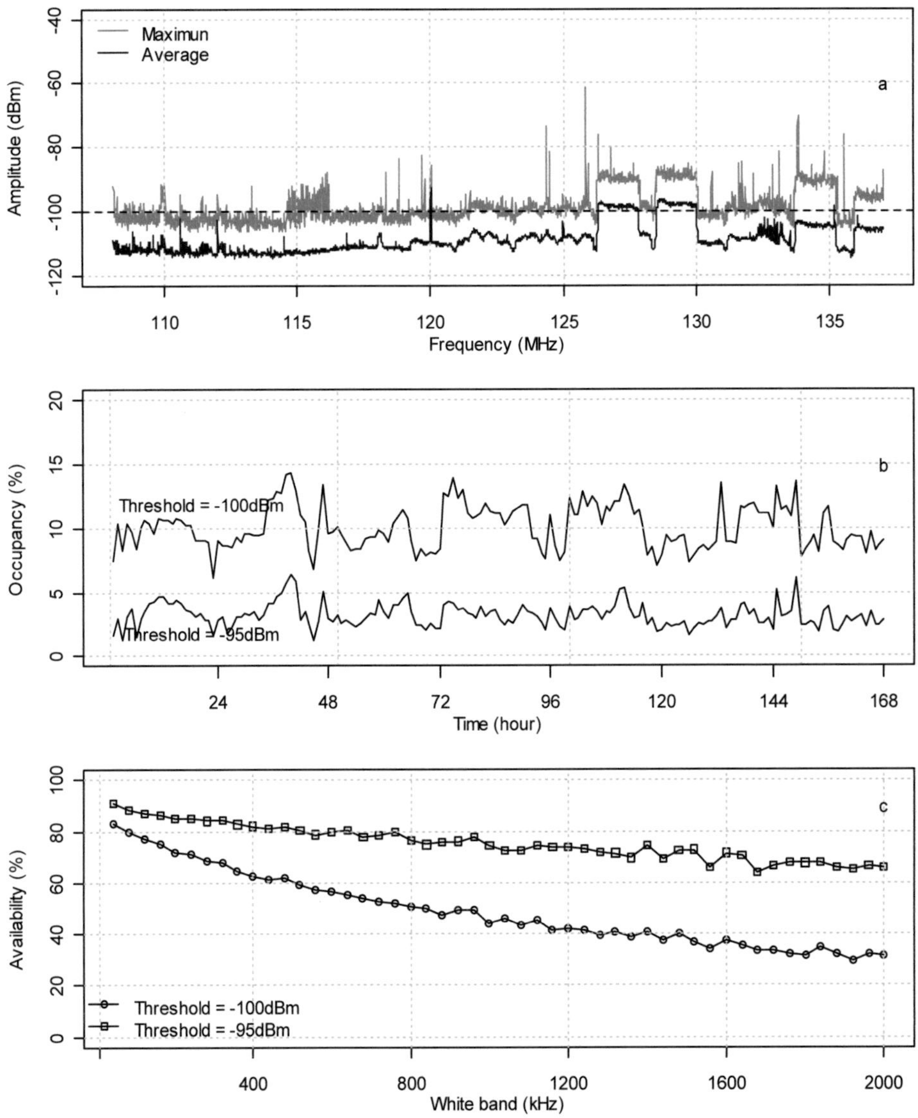

Air Band 108-137 MHz

	-105 dBm	-95 dBm
Min	6.06	1.20
1st Quantile	8.81	2.58
Median	9.76	3.03
Mean	10.06	3.16
3rd Quantile	11.19	3.67
Max	14.28	6.33

Fig. 6. Occupancy statistics of Air Band

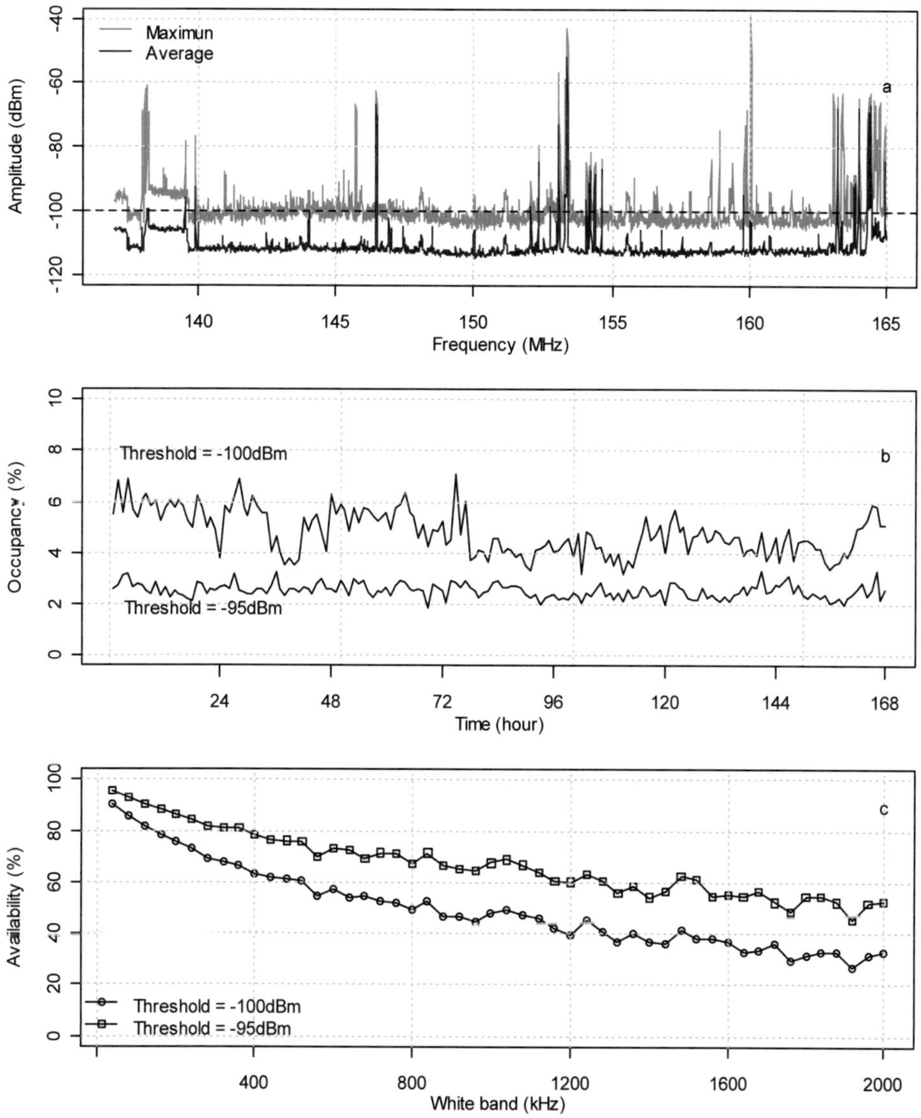

Mid Band 137-165 MHz

	-105 dBm	-95 dBm
Min	3.24	1.93
1st Quantile	4.13	2.35
Median	4.81	2.53
Mean	4.84	2.52
3rd Quantile	5.54	2.71
Max	7.09	3.32

Fig. 7. Occupancy statistics of Mid Band

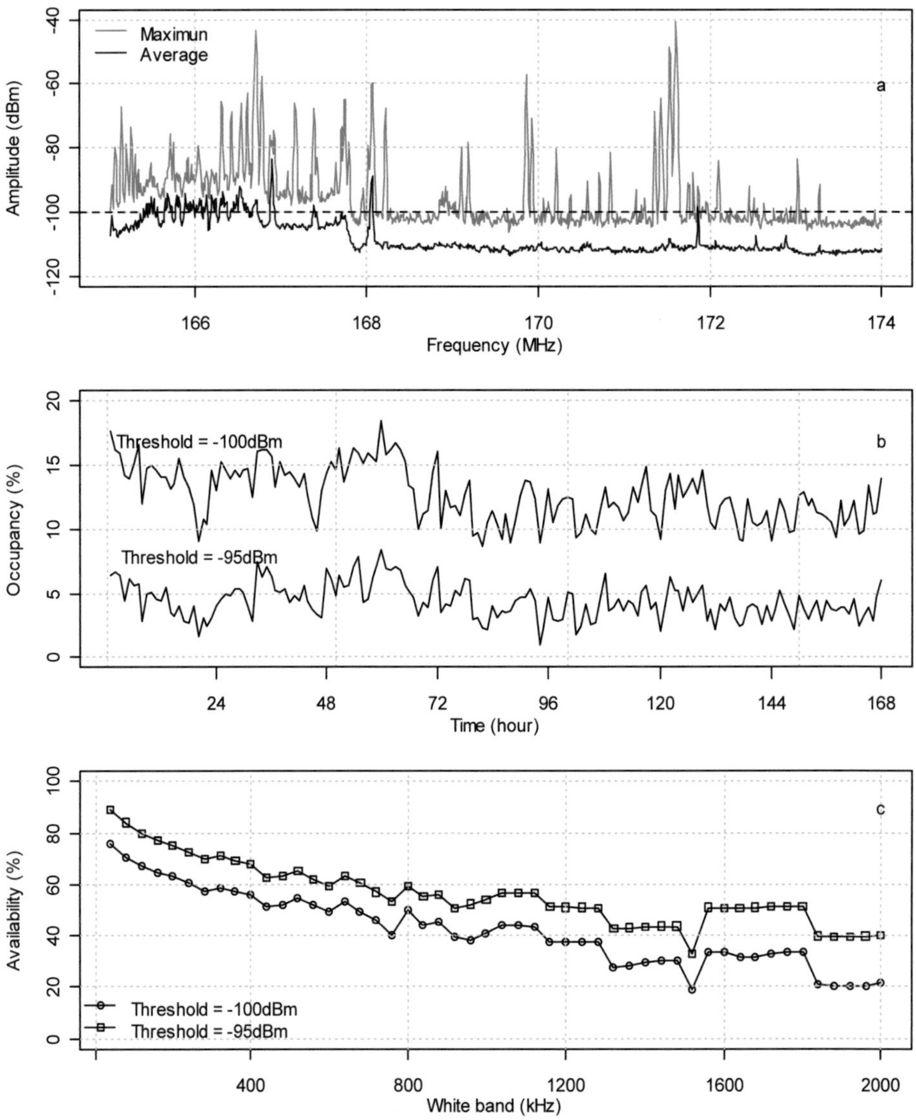

High Band 165-174 MHz

	-105 dBm	-95 dBm
Min	8.53	0.99
1st Quantile	10.74	3.30
Median	12.29	4.26
Mean	12.51	4.35
3rd Quantile	14.29	5.21
Max	17.94	8.31

Fig. 8. Occupancy statistics of High Band

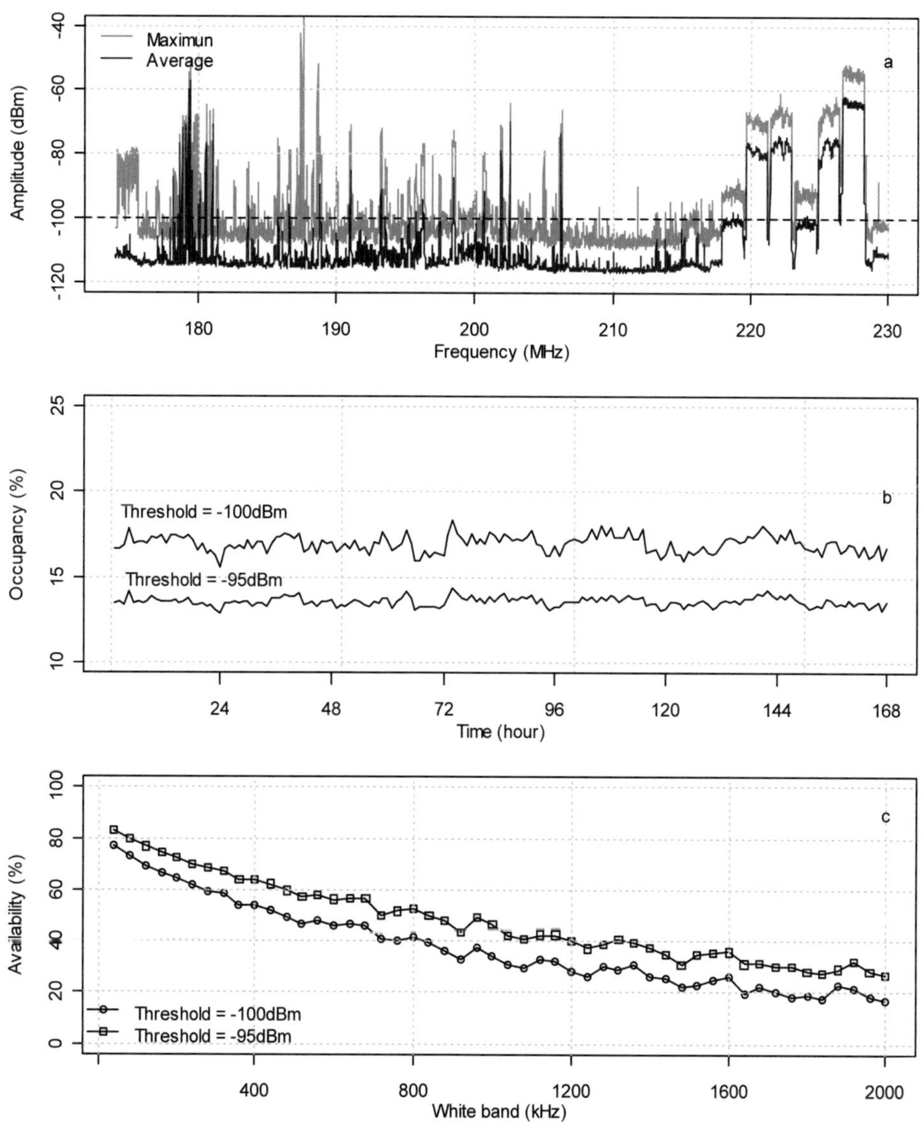

Band III 174-230 MHz

	-105 dBm	-95 dBm
Min	15.61	12.91
1st Quantile	16.61	13.42
Median	16.99	13.60
Mean	16.97	13.60
3rd Quantile	17.30	13.77
Max	18.36	14.37

Fig. 9. Occupancy statistics of Band III

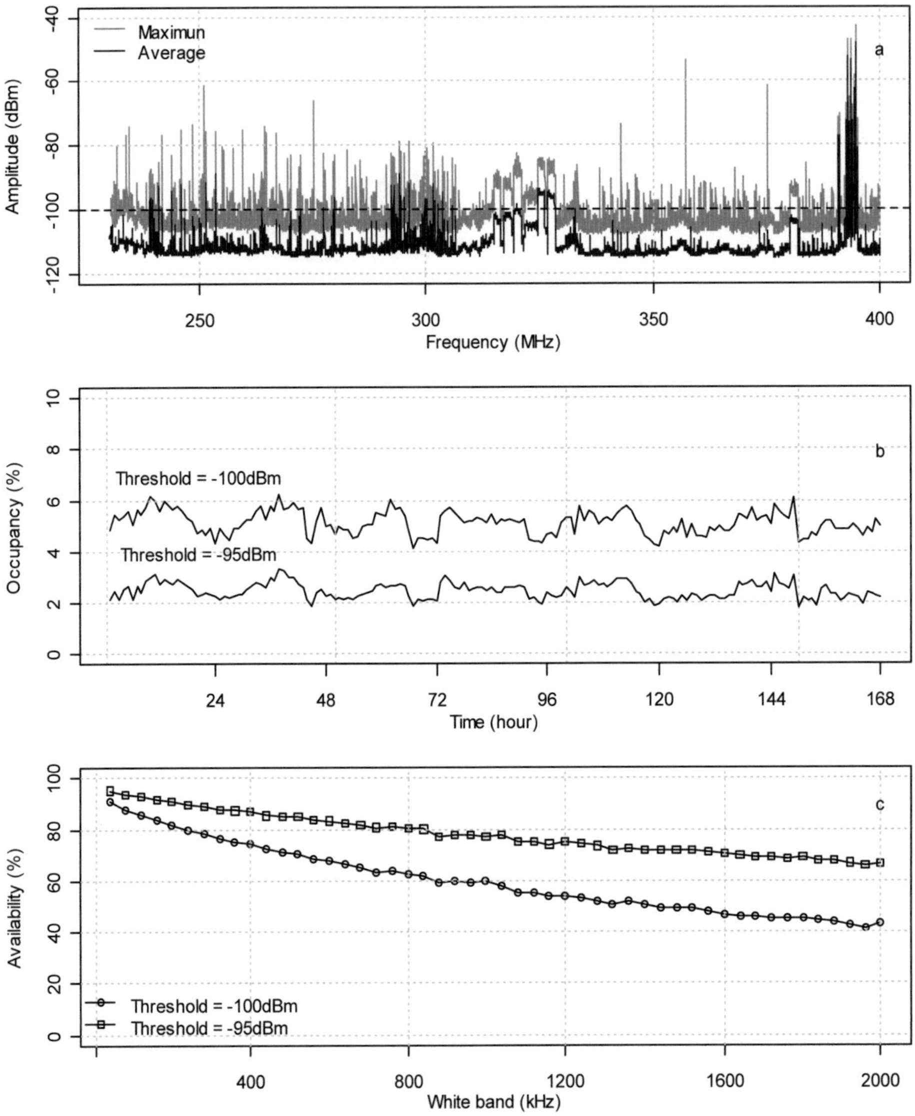

NATO Band 230-400 MHz

	-105 dBm	-95 dBm
Min	4.13	1.84
1st Quantile	4.77	2.20
Median	5.14	2.41
Mean	5.10	2.46
3rd Quantile	5.44	2.68
Max	6.19	3.29

Fig. 10. Occupancy statistics of NATO Band

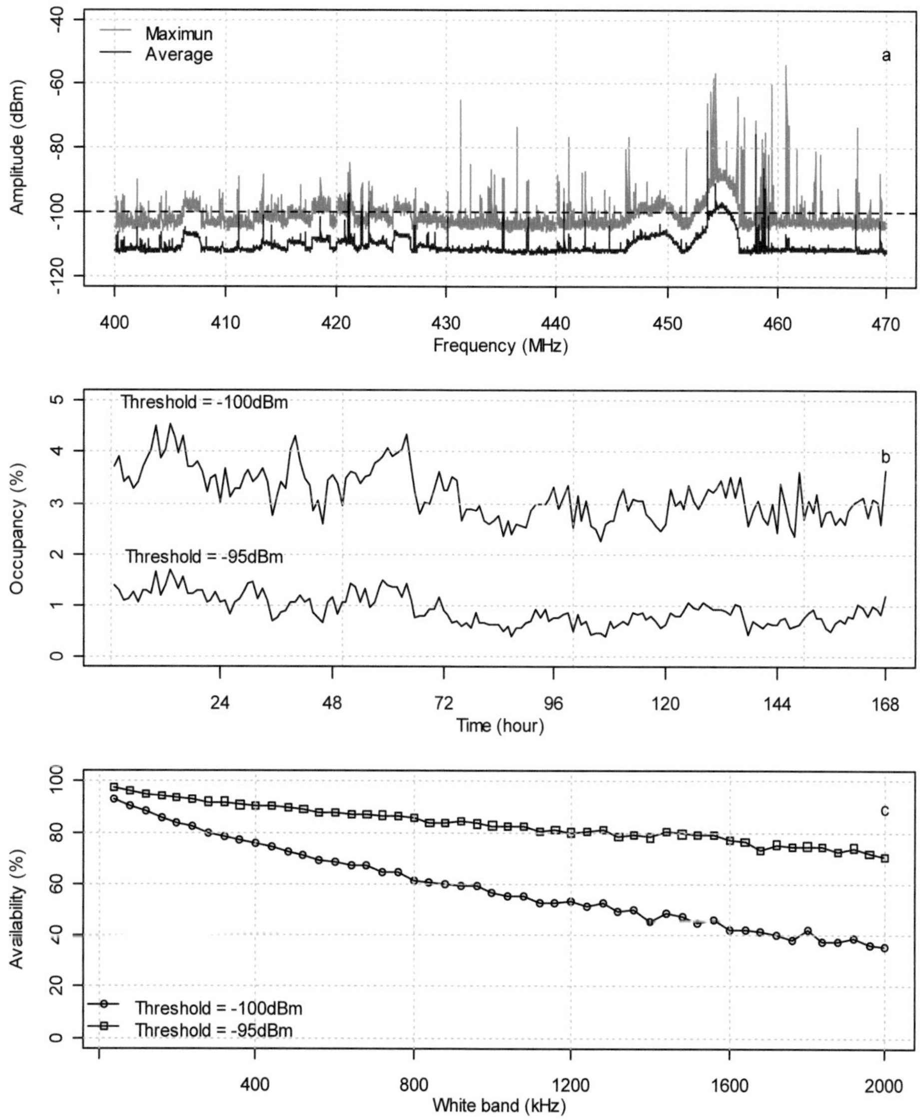

UHF 1 & 2 Band 400-470 MHz

	-105 dBm	-95 dBm
Min	2.25	0.40
1st Quantile	2.81	0.70
Median	3.12	0.90
Mean	3.18	0.93
3rd Quantile	3.51	1.11
Max	4.54	1.70

Fig. 11. Occupancy statistics of UHF 1 & 2 Band

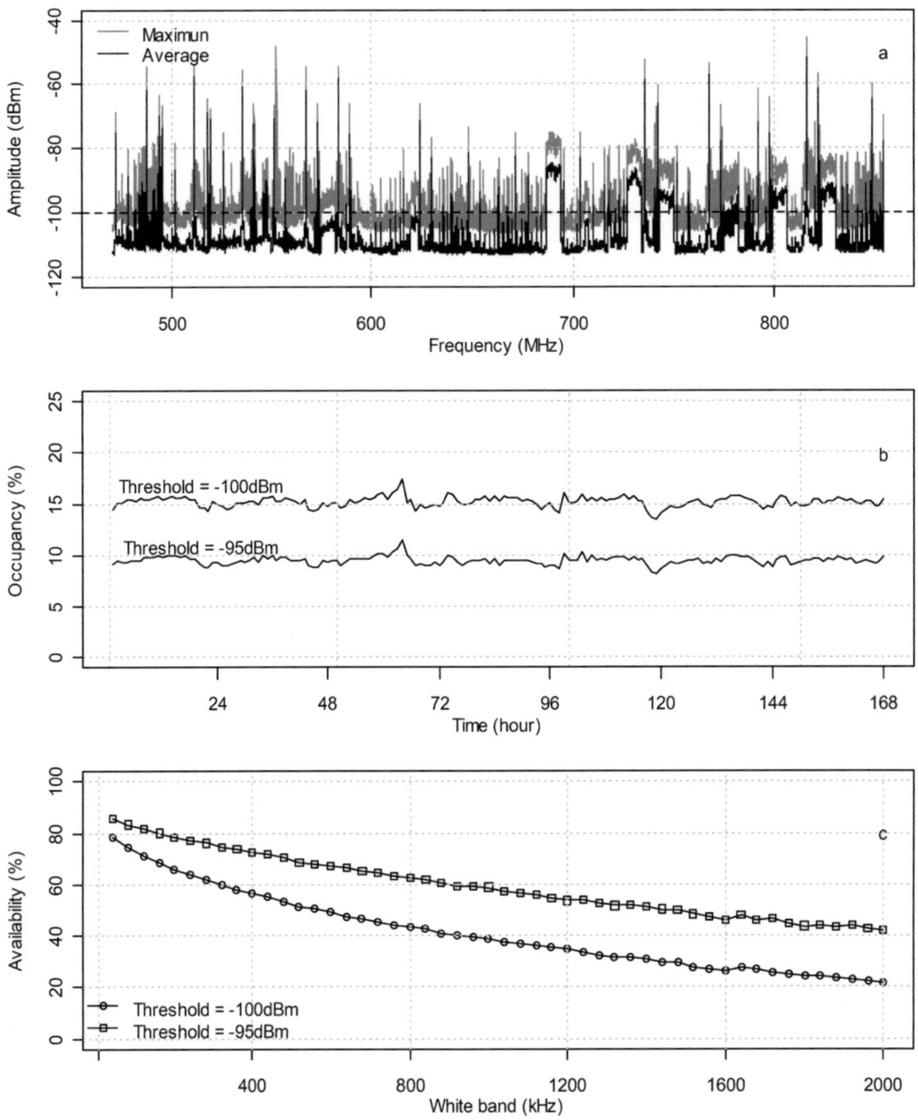

TV, Band IV & V 470-854 MHz

	-105 dBm	-95 dBm
Min	13.56	8.19
1st Quantile	14.95	9.26
Median	15.34	9.48
Mean	15.26	9.49
3rd Quantile	15.57	9.74
Max	17.51	11.58

Fig. 12. Occupancy statistics of TV, Band IV & V

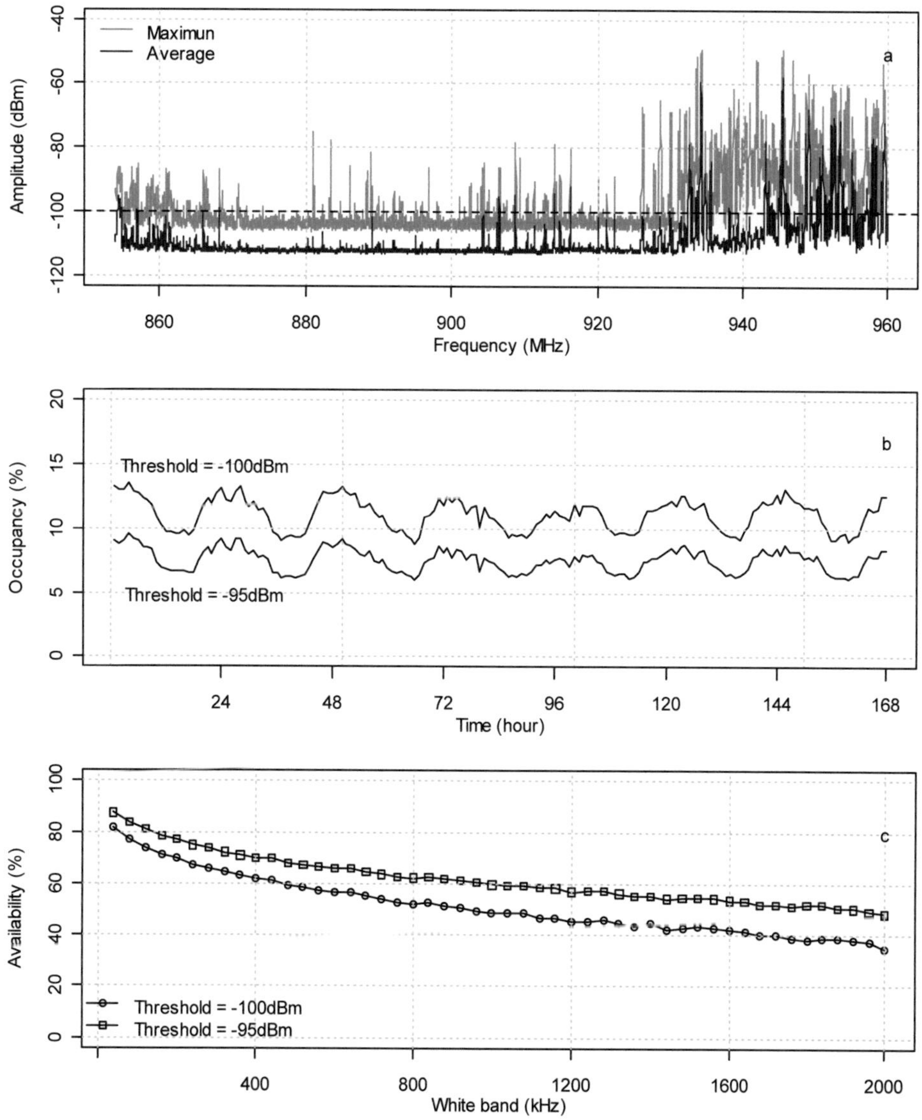

GSM & mobile 854-960 MHz

	-105 dBm	-95 dBm
Min	8.78	6.15
1st Quantile	9.74	6.72
Median	11.29	7.76
Mean	11.1	7.68
3rd Quantile	12.1	8.37
Max	13.65	9.80

Fig. 13. Occupancy statistics of GSM & mobile Band

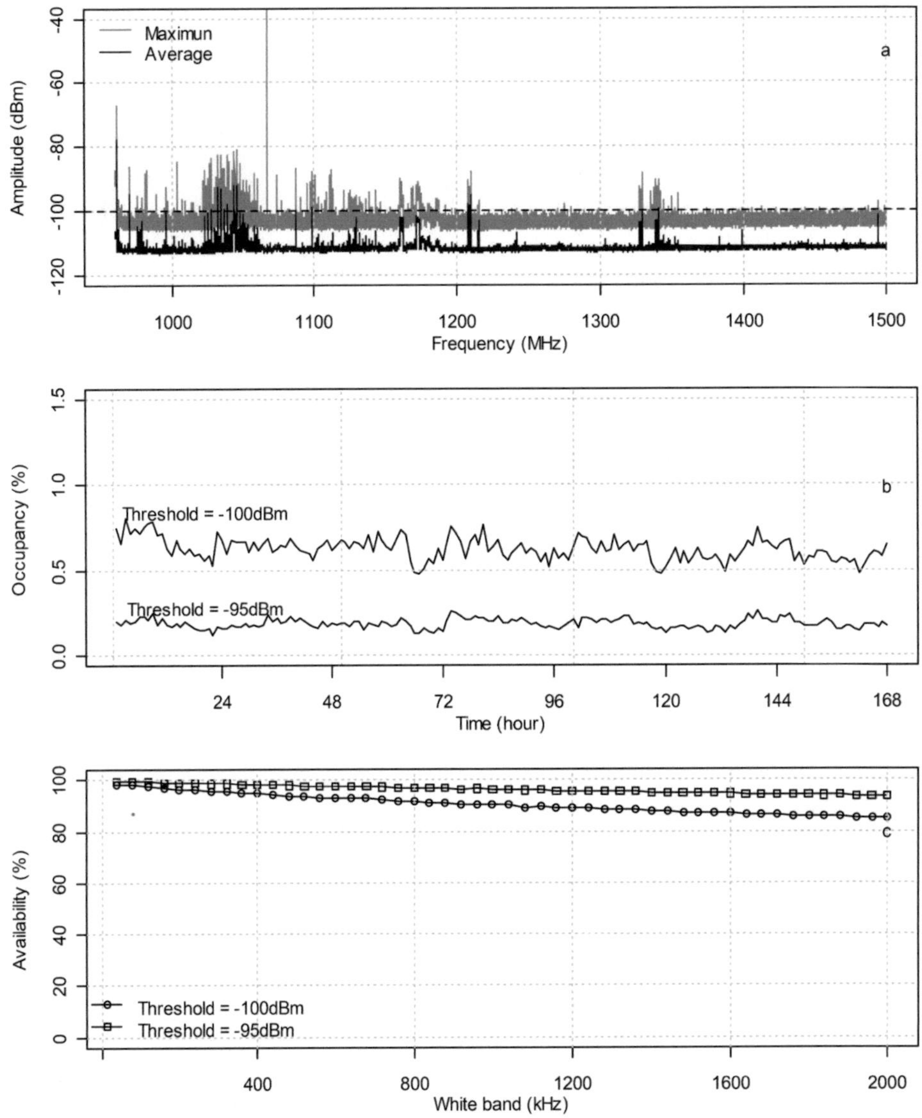

L Band (lower) 960-1500 MHz

	-105 dBm	-95 dBm
Min	0.40	0.06
1st Quantile	0.52	0.11
Median	0.56	0.13
Mean	0.56	0.14
3rd Quantile	0.59	0.16
Max	0.74	0.21

Fig. 14. Occupancy statistics of L (lower) Band

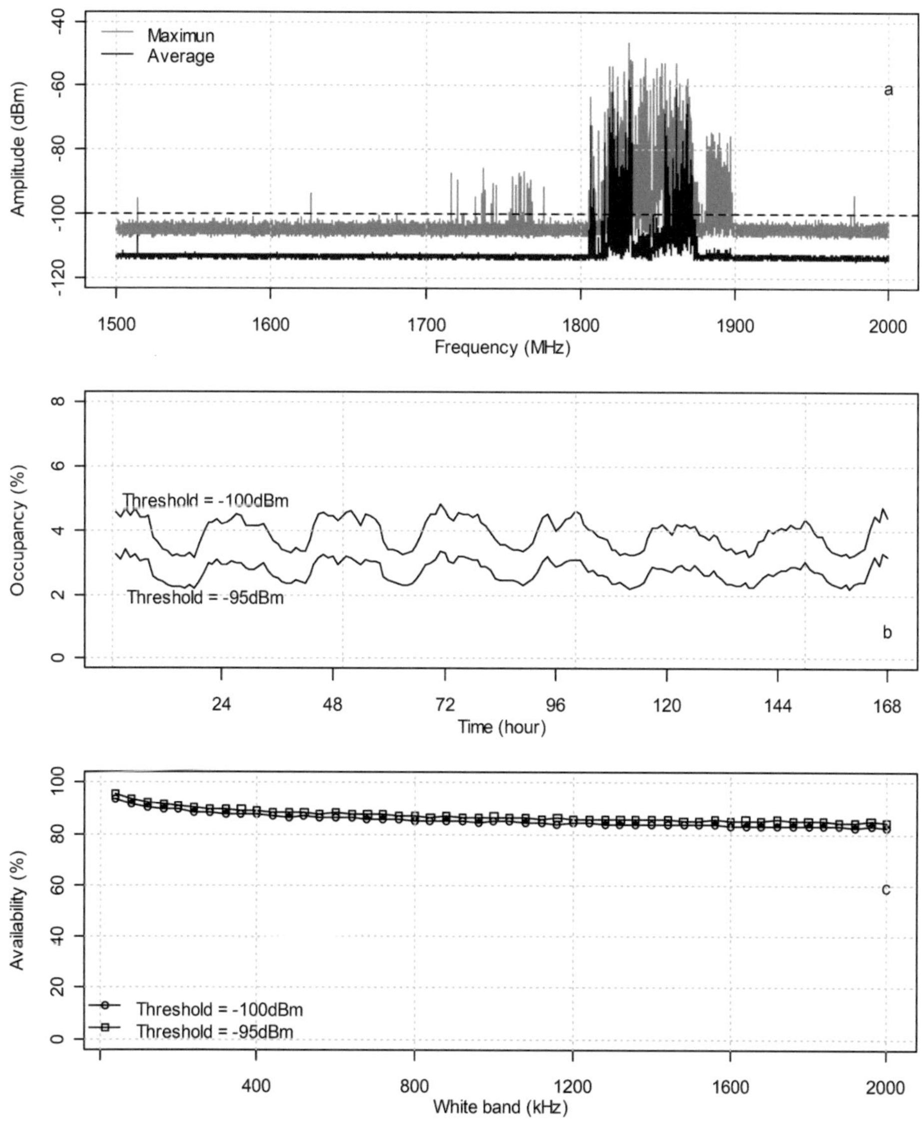

L Band (upper) 1500-2000 MHz

	-105 dBm	-95 dBm
Min	3.12	2.18
1st Quantile	3.43	2.40
Median	3.99	2.75
Mean	3.92	2.72
3rd Quantile	4.33	2.98
Max	4.82	3.34

Fig. 15. Occupancy statistics of L (upper) Band

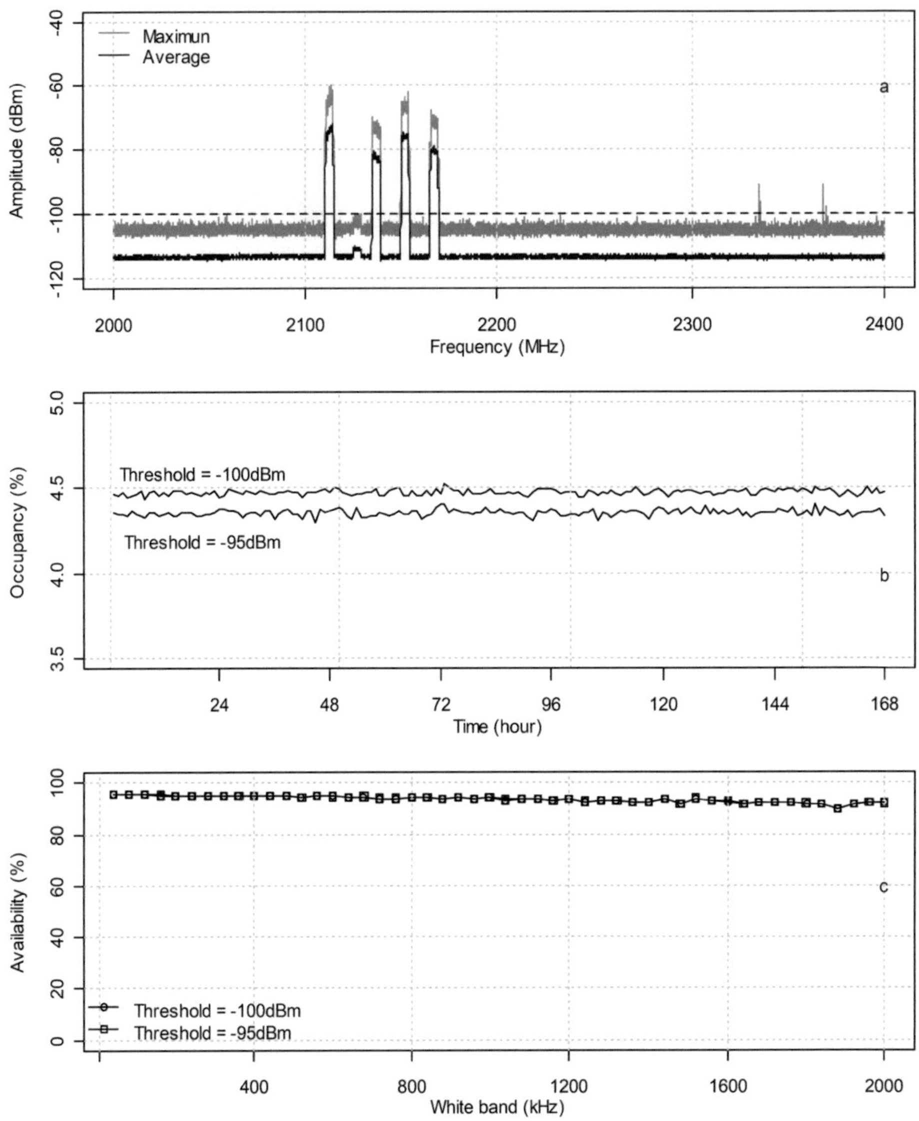

S Band (lower) 2000-2400 MHz

	-105 dBm	-95 dBm
Min	4.43	4.30
1st Quantile	4.46	4.34
Median	4.48	4.35
Mean	4.47	4.35
3rd Quantile	4.49	4.36
Max	4.52	4.41

Fig. 16. Occupancy statistics of S (lower) Band

The bottom panel shows the statistical distributions of the *white spectrum*. *White spectrum* can be defined as the continuous idle spectrum in a given bandwidth and in a given threshold which can be used for communications. For example, in Figure 12, for -100 dBm threshold we can find about 60% white band with 1000 kHz bandwidth distributed in the total spectrum, if we divide the TV band (470 - 854 MHz) into 384 sub-bands with 1000 kHz bandwidth.

The statistics table of each figure shows the minimum and maximum occupancy rates, 1st, 3rd quartile values, and mean and median values. A median is described as the number separating the higher half of a sample, a population, or a probability distribution, from the lower half.

The statistics presented in these figures shows that the spectrum occupancies are really spares in the measuring locations and period. The average occupancies for most bands are less than 20%. Except the GSM and CDMA communication bands, the average occupancies in the 1 GHz to 2.5 GHz spectrum are less than 5%. The good propagation characteristics of this range of spectrum, in terms of propagating distance and data rates, make it an excellent candidate for cognitive radio technology. The data shows that, without additional spectrum, there are a great amount of spectrum resources for accommodating the cognitive radio systems if the current communication regulations could be changed.

4. Conclusion

There are generally positive findings in this project with respect to the prospects for cognitive radio. Statistics show that the spectral occupancies are spare indeed. Occupancy rates of most bands in VHF and UHF are less than 10% overall and the distribution of the white band indicates that the bands are capable of running wideband wireless communications. Of course, the current static spectrum allocation policies and spectrum management strategies have to be modified to motivate applications of the cognitive radio technology. While this study is to identify the low utilizations of bands, long term studies are crucial in developing spectrum sharing technologies and for spectrum management.

Above all, radio spectrum occupancies in time, frequency and space domains observed in this project are sparse indeed. This suggests that cognitive radio technologies have great prospects in the future wireless communication infrastructure, if current telecommunication policies and regulations are modified.

5. Acknowledgement

I am indebted to Durham University and British Telecom for their funding of my studies. I would like to record my gratitude to Sana Salous for giving valuable guidance and suggestions for improving the work, and Stuart Feeney for sharing his RF knowledge, Roger Lewenz and Peter Baxendale for their contributions in the development of the data acquisition software.

6. References

Ashley, S. (2006). Cognitive radio, *Scientific American Magazine* pp. 35– 42.
Carr, J. J. (1999). *Practical Radio Frequency Test and Measurement*, Newnes, Oxford.
Freedman, D., Pisani, R. & Purves, R. (2007). *Statistical*, Oxford.
ITU (n.d.). *ICT regulation toolkit*, ITU.

Saakian, A. (2011). *Radio Wave Propagation Fundamentals*, Artech House.

Sanders, F., Ramsey, B. & Lawrence, V. (1996). *Broadband spectrum survey at San Diego, California*, Report, Department of Commerce, U.S.

Staple, G. & Werbach, K. (2004). The end of spectrum scarcity, *Spectrum, IEEE* 41(3): 48– 52.

Wang, Z. & Salous, S. (2009). Journal of signal processing system, *Spectrum occupancy statistics and time series models for cognitive radio* 62: 145– 155.